Four Hundred Years
of Agricultural Change
in the
Empire State

by Robert W. Bitz

LCCN: 2009910279
ISBN-13: 978-0-615-31865-3
First edition, published 2009.

Ward Bitz Publishing
Baldwinsville, NY

The author may be contacted at:
P.O. Box 302
Plainville, NY 13137

Cover photo: New West Washington Market in New York City from an 1888 photograph in Harper's Weekly.
Note that most of the produce is packed in wooden barrels, the dominant container at the time.

Acknowledgements

Many people have been helpful to me in writing this book and I am most appreciative of their assistance. Whoever I called upon was willing to help. Starting with the first chapter on Native American agriculture, Jane Mt Pleasant examined my draft and offered some suggestions that were very helpful. My friend Randy Nash, our New York State barn expert, read over the section on barns and gave me comments. Ed Rowley, Jessica Chittenden and Joan Kehoe all helped with the section on the New York State Department of Agriculture and Markets. Betsy Wright very kindly looked over my work concerning the Farm Service Agency. Bud Stanton read some of my material on mechanization and education, and gave me helpful suggestions.

I called upon some of my friends, who are actively involved in New York agriculture, to help me obtain information concerning their field of expertise. Jim Verbridge, Ken Pollard and George LaMont came to my aid with information about the fruit industry. Steve Wright and Jackie Arnold were most helpful with information concerning the fruit and vegetable processing industry. Tom Davenport helped me with the grape industry and George Mueller not only provided me information concerning the dairy industry, but also took me to the Wolf farm so I could see robots milking cows. My thanks to Mark Bitz who provided information regarding organic and natural.

Agricultural organizations have played an important part in New York's history and help was received regarding the Grange from Bruce Widger and Virginia Connor. Jeff Kirby gave me assistance regarding Farm Bureau and George Preston provided information concerning Cooperative Extension. Mark Stephenson looked over my material on dairy marketing and both Bob Lewis and Diane Eggert helped with information concerning farmers' markets.

I owe a bushel of gratitude to Debbie Stack who has gone over this book, chapter by chapter, correcting my phrasing, making comments like, "what do you mean by that?", and adding innumerable red marks so I could correct my punctuation. She also showed great restraint in letting my writing stand as presented, as I am sure she could have greatly improved the flow of words. It has been a long process and Debbie's help has been valuable.

As part of the acknowledgements I want to recognize my parents, Harry and Metta Bitz, both long deceased, who taught their son the joy of work, especially work well done. This book has been a great deal of work, but as I often say, "If you enjoy what you do, you never have to work." I have enjoyed writing this book. I also want to acknowledge Stan Warren, my Professor of Farm Management at Cornell back in 1949. Stan was an outstanding teacher and I cannot but believe he helped me with this book.

To all the aforementioned people I extend my gratitude. I apologize to those that I inadvertently overlooked.

About the Author

Robert Bitz was the fifth generation of his family to operate the family farm in Onondaga County. The farm, originally 70 acres, was purchased in 1835. Growing up, Robert worked with a dairy, turkeys, horses, tobacco, sweet corn, cabbage, potatoes, beans and a variety of other crops. It was typical of the thousands of other New York farms until it specialized in turkey production in the 1950s. He has seen dramatic changes in both the State and its agriculture.

Bob received a B.S. at Cornell University and returned to the family farm specializing in turkey production and marketing. He built the turkey business into one of the largest in the Northeast, marketing the farm's turkey products in numerous states. He also has been active in community affairs serving on school, library, planning and town boards. He has been on the Federal Reserve Bank's Agriculture and Small Business Advisory Board, several bank boards and Cornell's CALS and Veterinary School Advisory Boards. He has been involved with Cornell Cooperative Extension, Farm Bureau and in the original development of Agricultural Districts. He has served as a Cornell Trustee and on the Board of Directors of the Witter Agricultural Museum and the New York State Agricultural Society.

Travels have taken him throughout New York State and to many countries, including both developed and developing, where he has seen agriculture practices differing by hundreds of years. He enjoys history and for a number of years had a Museum on his farm, called "The Pioneer Experience". He has a substantial collection of antiques relating to pioneer and farm life. One of his hobbies is writing. He recently wrote, *A History of Agriculture in Onondaga County*. Although he has passed the family farm on to his son, Mark, he maintains his avid interest in agriculture.

Preface

Although agriculture produces the food we consume on a daily basis and most of us wear clothing made from natural fibers, only a few people give thought to how these items were produced, where they came from and how they reached the market place. We give even less thought to the changes that have taken place in New York agriculture since Henry Hudson sailed up what is now called the Hudson River, in 1609. Much more has happened than can be related in one book, but New York's agriculture has been an extremely important part of our State's history and I hope the material in this book will help those who read it gain a better understanding of the tremendous changes that have transpired during these past 400 years.

Another reason for writing this book is my lifelong love of agriculture. For generations my family has been farmers and I never gave serious thought to any vocation other than agriculture. As a boy, my life was centered on farming, since almost all of our neighbors, friends and relatives were farmers. These neighboring farmers helped each other without being asked, whenever there was need. The community acted as one large family.

For me, an appealing aspect of farming was the great variety of work it encompassed. During the 1930s-40s the majority of farms were still general farms growing a variety of crops and livestock, which provided a form of "self insurance". If one or two crops failed there were usually others that did well, adverting a year of total failure. There was little chance for boredom, as seldom would the same task be repeated for more than a few days. Each crop had its own timetable for planting and harvest and each kind of livestock and poultry had its own special requirements. Our farm milked 15 Guernsey cows, grew 3,000 turkeys, and in addition to the normal hay, wheat, corn and oats, grew cabbage, tobacco, potatoes and red kidney beans. There were only a few acres of each crop but because of the additional labor growing and processing the turkeys, my dad had two hired men.

Five horses powered our farm until 1939, when the purchase of a Farmall F-20 became our first major step toward mechanization. The horses stayed fully employed for several more years, but by the mid-40s they gradually spent more of their days in the pasture as tractors replaced them. With the tractors came larger equipment, requiring an increase in acreage and a growing movement toward specialization to gain greater utilization of specialized equipment that resulted in greater efficiency.

I feel fortunate to have been born in the era of the horse-powered general farm, but also at a time to participate in the rapid mechanization in agriculture. My boyhood was not appreciably different than that of my father or grandfather, but was totally different than the boyhood of my sons and especially, of my grandsons. The changes in agriculture that I have been fortunate to experience provided another incentive in writing this book: to document some of these changes. Rather than tell many of my personal experiences, I have chosen to show some of the factors that created the changes affecting the lives of hundreds of thousands of New York State farm families during the past several hundred years. Each farm family has been affected by change in slightly different ways. Some discontinued farming by choice, others by necessity. Some chose to adapt new techniques too early and others too late. Some were born at the right time to experience good prices whereas others were born at the wrong time and started farming in a period of low prices. I marvel, though, at how many farm families have persevered through low prices, and weather or insect related crop failures.

The changes in the agriculture of New York have been greater than most of the other states. When the white man arrived the agriculture of the Iroquois was already well-developed, producing quantities beyond their needs. Settlement by Europeans brought agriculture practices in the tradition of both the Dutch and the English. Soon after the Revolutionary War, New York became the number one agricultural production state in the country, holding this distinction for many years. Natural waterways to both the north and the south provided routes to export agricultural products, even prior to the construction of the Erie Canal. With the Erie Canal completed, New York became the passageway to markets for a large area of the country and appropriately earned the label, "The Empire State". Railroads soon followed the route of the Erie Canal, connecting New York State and its New York City harbor to a greater portion of our country, and providing the means for shipment of products throughout the world.

I suggest that the reader take some time to look through the **Appendix** of this book. There are numerous tables that compare New York's agriculture with that of the other states in the middle of the 19th century. These tables show the dominance of New York's agriculture at that period of time. There are also tables in the Appendix that compre the agricultural production of the various counties of the state. In addition, there is an alphabetical list of the counties showing county seats, date created, entity from which they were formed, source of their name, their population and size. All of this information will provide the reader with a better understanding of our state and its agricultural heritage.

It is impossible to include in this book all of the important changes that have occurred in New York's agriculture. Of the things left out, some are intentional but others are oversight. In addition, what seems important to one person may not be to another. I believe figures often can tell more than words, so I have included tables and numbers to help provide a picture of what was happening at a specific time and the changes made over time. New York published a census, which included some agricultural information, in 10-year increments from 1825

to 1875. The U.S. Census Bureau has published an agricultural census since 1850, normally in 10-year increments, but occasionally in shorter increments. The New York State Department of Agriculture and Markets has published a book of Agricultural Statistics for a number of years. All of these sources have been a great help in determining what has been happening in New York agriculture during the past 185 years.

A History of Agriculture in the State of New York, written by Ulysses Prentiss Hedrick and printed for the New York State Agricultural Society in 1933, has been a most valuable source of information for this book. There are numerous pages where I refer to information from this book. Another book that is a joy to read and that has been helpful is, *The Golden Age of Homespun,* by Jared van Wagenen, Jr., published in 1953.

There are over 100 pictures included in this book to help the reader understand agriculture more effectively. There are a few pictures of today's agriculture, to help the non-agriculturally informed reader and readers of the future better understand the agriculture of today. It is unfortunate that there are not pictures of the first 250 years of our agriculture available, but a few drawings of those early years are included.

I cannot emphasize enough the great help that Cornell University, with its programs of research, teaching and extension, has provided to New York agriculture. Since the late 1800s, the scientific knowledge and general information provided to our farmers by Cornell has been immeasurable. New York State government must not be overlooked, as the State has provided substantial funding for the betterment of agriculture and the Department of Agriculture and Markets offers continuing support.

I feel guilty for not mentioning more of the great leaders of New York agriculture, but that is an impossible task. There have been thousands that have not only tilled the land, but have spent innumerable hours working for the betterment of our agricultural industry. In my limited span of only three quarters of a century, I have seen three generations of great leaders step up and give of their time and resources for the betterment of our agriculture. Some are worthy of mention because of their on-the-farm innovations, while others have been researchers, teachers or simply great individuals because of the example they set. If I were to name 100 of these wonderful people, I would miss 100 and if I were to name 1,000, I would be overlooking another 1,000.

As I place praise in the direction of the hundreds and thousands of agricultural leaders that we have had in this State, I also want to emphasize that their leadership has not only helped agriculture, but has heaped untold benefit upon the non-agricultural residents of the State. Whenever there is an improvement in agriculture the benefits soon reach the consumer through improved food, lower cost, or both. Agriculture prospers from innovation for only a short time, with the long-term benefits passing on to the consumer.

For many generations, farming was a way of life as well as a business, but this too, has been changing during the past 70 years. Innovations in transportation and communication have brought tremendous change, not only to the farm, but also to life on the farm. Both family life and farm children have gradually become more like their counterparts in suburbia and the city. Television, movies and the Internet, as well as rapid transportation to and from the centers of population, are providing similar influences wherever people live. Today, our farm children have the same opportunities as other children and are not expected to come back to the farm unless it is their desire. Often the older generation discourages their return to the farm because of the opportunities that exist in other careers. There may be a counteraction to this in

progress currently. An increased number of people are entering agriculture that do not have an agricultural background but because of their concern for food and environmental safety. In any event, unless the US government changes its policy of cheap food or consumers become willing to pay more for their food, the number of commercial farms in New York will continue to decrease.

There will continue to be small family farms in New York for generations, but they will often serve as rural residences with an "off-the-farm" full-time job. Existing family farms will become larger and fewer as competition and the call of "off-the-farm" occupations increase. There will be an increase in large corporate farming, but New York's diverse agricultural production and many areas of small pockets of farmland will not attract large corporate agriculture as much as some of the other states that are more homogenous in their production and with larger expanses of fine agricultural land.

Throughout our history only the very best have been able to survive in farming. Today, it is increasingly difficult to survive in agriculture, with competition coming from all corners of the world. To be successful, a farmer has to be astute in business, be extremely productive and very efficient. Except for niche and specialty markets, the farmer has to keep costs low to make a profit. Farming, as a way of life, still continues on a few farms but has been rapidly disappearing. It is difficult for me to see these changes, but change is inevitable, and this book is a story of change.

Table of Contents

Table of Tables

Chapter I
Native American Agriculture

"The quantity of corn destroyed, at a moderate computation, must amount to 160,000 bushels, with a vast quantity of vegetables of every kind," stated General Sullivan on September 30, 1779 in his report of the "Campaign to Destroy the Iroquois". This report was directed to John Jay, President of the Continental Congress. The report was lengthy and mentioned many of the expedition's other activities including the destruction of 40 Iroquois towns and scattered Iroquois houses, as well as hundreds of acres of corn and an orchard of 1,500 fruit trees.[1] Although unintended, this report provides us with valuable information about Native American agriculture.

For those unfamiliar with a quantity of 160,000 bushels of corn, it represents the equivalent of 100 of today's tractor-trailer loads of corn! Dr. William M. Beauchamp estimated the total Iroquois population at the beginning of the Revolutionary War at 15,000.[2] He also estimated the number may have been as high as 25,000 before the settlement of the white man. Whatever the actual number, there is no question the Iroquois were active agriculturalists.

General George Washington sent General Sullivan to destroy the Iroquois for two reasons. The Iroquois had made numerous attacks upon white settlements in the Mohawk Valley killing many settlers and destroying their property. It was estimated that one third of the white residents of the Mohawk Valley were killed during the war and another one third fled for their safety.[3] The Iroquois had some justification for these attacks as white colonists had continuously encroached on Iroquois lands in violation of established treaties between Great Britain and the Iroquois Confederacy. In addition, during the Revolutionary War, the Iroquois were helping to supply the British Army with food. Several thousand Colonial soldiers participated in this offensive and many were so impressed with the fertile land in Central and Western New York that they settled there after the war.

This Colonial Army offensive came more than 150 years after the first settlement by the white man in New York and more than 250 years after Giovanni da Verrazano, sailing for France, brought the Dauphine into lower New York Bay.[4] The report of Sullivan's Campaign gives us a small picture of native agriculture in 1779, but what was it like before the entrance of the white man into his life? Archeologists have unearthed a great variety of native artifacts dating thousands of years before the arrival of the white man. These artifacts, consisting of stone tools, stone weapons and pieces of pottery and bone, tell us a bit about Native American life but provide only a few clues as to the extent of their agriculture. These remnants tell us that agriculture existed but do not tell us either its length or extent. Early explorers provide us with more extensive information from their firsthand accounts.

Henry Hudson, sailing for the Dutch East India Company in 1609, traveled up the Hudson River as

1   The Sullivan-Clinton Campaign in 1779 (State University of the State of New York 1929)
2   Halsey, Francis Whiting *The Old New York Frontier*
3   Roberts, Ellis H. *American Commonwealths* p. 450
4   Kammen, Michael *Colonial New York* p. 1

far as it was navigable, a little north of Albany. He made detailed records of his trip and noted that near what is now Catskill, natives traded ears of corn, pumpkins and tobacco for the trifles of the sailors. Further up the Hudson, natives came on board with grapes, pumpkins and beaver and otter skins, which were traded for beads, knives and hatchets.[5]

There were numerous Native American tribes in what is now New York when Henry Hudson made his voyage up the Hudson. There were tribes on Long Island, along the Hudson River and in other parts of the State but none were as numerous, strong or feared as the Iroquois. The Iroquois controlled the land north and west of the Mohawk River. Five tribes: the Mohawk, Oneida, Onondaga, Cayuga and the Seneca, had formed an alliance to protect each other against invading tribes many years before the white man arrived. In 1715 the Tuscaroras became the sixth tribe in the Iroquois Confederacy. This defensive alliance also turned into an alliance for aggression, as the Iroquois became feared by neighboring native tribes as well as tribes in Canada and to the south of New York. John Smith, of the Jamestown, Virginia settlement, recorded that in 1608, he encountered an Iroquois war party on the Chesapeake. The Iroquois even obtained tribute from the native tribes on Long Island in the form of taxes.[6]

In western New York, other than at Fort Erie and Fort Oswego, there had been a limited amount of interaction between whites and natives until the Revolutionary War. Explorers and fur traders had visited and passed through the area numerous times but no settlers had ventured into the land, deterred both by its remoteness and fear of the natives. Prior to contact with the white man, the natives had only stone, bone and wood tools and no firearms. Upon seeing the advantages of iron tools and firearms, the natives eagerly traded furs for these items. They also obtained apples, peaches and other fruit from white men for planting orchards. The seeds for the fruit trees probably came from the French Jesuit missionaries who came to central and western New York from Canada. Previously, the only fruit cultivated by the Iroquois was the black plum. The natives had adopted the white man's way of using domesticated livestock since, when Sullivan's army

came to destroy the natives, the soldiers observed domesticated cattle and horses running from Indian villages.

We know from many sources that the only domesticated animal the natives possessed, prior to contact with the white man, was the dog. It served as a pet, warned of intruders and scared birds from the fields of corn. Dog meat also was used for food on special occasions. The Green Corn Festival was a period of thanksgiving celebrated by the Iroquois and at its end there was a sumptuous feast of "succotash". Succotash was prepared in several different ways but was at its best when the well-chopped fat meat of a dog was added to a mixture of corn, beans and squash.[7] The dog was most certainly the Native American's best friend!

The Native Americans were excellent hunters and wild game provided a significant portion of their diet. They were also good fisherman with fish especially important to tribes living near water. Sometimes weirs of stone were constructed in rivers to trap fish and hold them until the need arose for their consumption. The native tribes along the Hudson and on Long Island did not till the soil as much as the Iroquois. The Long Island natives' diet ran heavily to oysters and fish.

We have often read of the "Three Sisters" grown by the natives. These were corn, beans and squash that were grown together in fields. There was a synergism by growing these together. The corn supported the vines of both the beans and the squash and the squash shaded the ground helping to prevent the growth of weeds. The beans supplied nitrogen to the soil and were an important source of protein in the diet. The natives had two major varieties of corn, white dent and white flint, with an occasional red ear. They also had three, less used varieties: popcorn, sweet corn and a starchy sweet corn.[8] There were several varieties of beans, all Phaseolus vulgaris, usually a climbing plant. There were several varieties of summer squash, pumpkins and gourds. They also grew varieties of winter squash and pumpkins but it is unknown if these were grown before contact with the white man.

5   Roberts, Ellis H. *American Commonwealths, New York*
     p. 21-22
6   Ibid. p. 129-134

7   Hedrick, Ulysses Prentiss *A History of Agriculture in the State of New York* p. 24
8   Onion, Daniel K. *Corn in the Culture of the Mohawk Iroquois* p. 60

The "Three Sisters" mound system of cropping is worthy of close scrutiny when we consider our agricultural practices of the past century and the current trend toward sustainable agriculture. This system utilized by the Native Americans was a sustainable agriculture for several hundred years, using minimal tillage and visual selection for improving varieties. Their successful agriculture was accomplished without inorganic fertilizers and pesticides. The crop yields were at least equal to those of the white man on comparable soils in the same period. They were not only able to produce corn grain yields of from 25 to 75 bushels for generations, but produced beans and squash at the same time from the same acres. In addition, there was minimal erosion and soil compaction and plant residue was returned to the mounds enriching them with organic matter.[9]

Ulysses Prentiss Hedrick, in *A History of Agriculture in the State of New York*, lists other plants cultivated by the natives including the Jerusalem artichoke, gourds, groundnut, leeks, pumpkins, and tobacco. He also lists a great variety of native plants used as food but apparently not cultivated, some of which were blackberries, beechnuts, blueberries, chestnuts, cherries, cranberries, currants, elderberries, gooseberries, grapes, huckleberries, turnip, plum, raspberries, strawberries, wild rice and wintergreen. Some plants used for other than food were bayberry for candles, bloodroot for dye, bulrushes and cattails for mats and baskets, dogwood for smoking, grasses for mats, hemp for fiber and indigo for dye.[10]

In Native American society each gender had its own roles, roles that differed from those of white men and women. The Native American woman was the farmer. It has been said, "the woman was the native's mule". The men helped clear the land of trees but since work was divided by gender, they felt it wasn't manly to work with the crops. It was woman's work to set a fire around a tree that had been girdled and thoroughly dried. The fire brought the tree to the ground often burning much of the stump in the process. She would then, with her stone or wood hoe and wooden spade, pull up the small roots and prepare the ground for planting.

The female Native Americans were the "Three Sisters" farmers, caring for the crops from before the seeds were planted through harvest. The corn seeds were soaked, in a solution distasteful to crows, to begin germination before planting. For fashioning a hoe, the bone of the shoulder blade of a bear or moose would be fastened to a wooden handle or for natives living near the seashore, a clamshell. The ground was mounded into hills that were three or four feet apart and several kernels of corn were dropped in a hole made with a small round stick. Later, bean seeds were planted in the same hill to provide the corn a head start and squash was planted in between the rows. Scarecrows were used to scare birds away and snares were set to catch the birds that ignored the scarecrow. A snared bird would be hung from a pole to help scare other birds away.[11]

The woman did the hoeing, weeding, scared birds away, harvested the crops and prepared the meals. August was a joyous time when ears of green corn were ready to be boiled or roasted in the husks. Some was eaten directly from the cob and some was boiled, parched over coals, husked, dried in the sun and then stripped from the cob. This parched corn would keep a long time and was often taken by the natives as food when traveling. The majority of the corn was harvested in the fall. When the corn was husked, four husks were sometimes left on the ears for braiding together into three-foot wide strips for hanging to dry. The largest ears were saved for seed corn, which was also left on the cob.[12]

Some of the harvested corn was dried on mats, care being taken to cover it at night with other mats and then to uncover it when the sun was shining. When thoroughly dried, it was husked and stored in caches, lined with dried grasses and covered with bark, in the ground. Some was stored in wooden receptacles about three feet high, made by cutting hollow logs into sections or was stored in baskets in the home. It was common for the women of a family to produce from 24 to 60 bushels of unshelled corn dried this way. An estimate of the acreage of corn grown by a single family was between half an acre

9   Mt. Pleasant, Jane *The Science behind the Three Sisters Mound System* p. 529-536

10   Hedrick, Ulysses Prentiss *A History of Agriculture in the State of New York* p. 31-33

11   Onion, Daniel K. *Corn in the Culture of the Mohawk Iroquois* p. 61

12   Ibid. p. 61-66

and one and one half acres. With a village of 100 families, this could amount to 100 acres.[13]

Fertilizer was not commonly used except where a large supply of fish was available. In this case, a fish or two might be added to a hill of corn. After crop harvest, corn stalks, along with the bean and squash vines, were incorporated into the hills to provide mulch and fertilizer for the following year's crop. The land was farmed annually except when the soil fertility was replenished by a year of fallowing. Eventually, a new area was cleared for crop production when the village moved to a new location.

Tobacco, a smaller variety than is common today, was grown by some of the tribes. It was the only crop commonly grown by the men and was one of the items traded among the tribes along with corn, shells for making wampum and arrowheads. Not all tribes had a flint quarry in their territory so there was substantial production and trade in flint tools and arrowheads. When they gained access to the white man's rifle the native arrowhead industry was decimated. This may have been the first American industry driven out of business by a superior product.[14]

The maple tree, scattered through much of New York, was a source of maple sugar for the native tribes. The native production method is described in a 1753 published account by Rev. Samuel Hopkins of Springfield, Massachusetts:

> *"The Indians make their sugar of the sap of maple trees. They extract the sap by cutting the tree on one side in such a form as that the sap will naturally gather into a small channel at the bottom of the hole cut; where they fix into the tree a small chip of six to eight inches long which carries the sap off from the tree, into a vessel set to receive it. Thus they tap a number of trees; and, when the vessels are full, they gather the sap and boil it to such a degree of consistence as to make sugar. After it is boiled, they take it off the fire, and stir it until it is cold, which is their way of graining it."*[15]

Other lesser uses of maple sap were to brew a weak beer and to make vinegar. To make these they boiled down three parts of sap to approximately one part vinegar or beer.

There is a long list of native agricultural processes and innovations prior to contact with white men. They lived in settled villages and cultivated the soil. They prepared their land and planted it systematically. They controlled the weeds and used scarecrows to keep away birds. In addition, they developed the husking peg, domesticated and cultivated wild plants, practiced plant breeding by seed selection, practiced multiple cropping, cleared forests by girdling and fire, made maple syrup and sugar by evaporating sap, and preserved fruits and berries with syrup, sugar and honey. They also preserved meats by drying and smoking, protected vegetables from rotting bacteria and fungi by burying them in the ground, extracted oil from nuts by boiling them in water, extracted paints, dyes and stains from plants, utilized vegetable fibers by spinning and weaving and developed agriculture to such a degree that their subsistence was chiefly from cultivated plants. A major contribution to today's agriculture was the development and cultivation of corn. It was estimated that Native Americans, throughout what is now the United States, grew as much as one million bushels of corn a year with a substantial portion grown in what is now New York.[16]

The farming methods utilized by the native tribes of New York may seem crude and primitive by the standards of today. But in reality, they are deserving of respect. Before any contact with the white man, and without a written language or beasts of burden, they were living well on the fruits of their successful agriculture. Jesuit priests, who came into New York from Canada in the early 1600s, called the Genesee country, "the Indian Granary". Western New York had a good reputation for its agriculture before any white person had made settlement. Unquestionably, Native Americans were New York's first farmers.[17]

13 Carver, T. N. *Historical Sketch of American Agriculture* in *Bailey's Cyclopedia IV* p. 31-32

14 Hedrick, Ulysses Prentiss *A History of Agriculture* p. 36

15 Ibid. p. 35

16 Holmes, G. K. *Bailey's Cyclopedia of American Agriculture IV* p. 37-38

17 Hedrick, Ulysses Prentiss *A History of Agriculture* p. 38

Chapter II
Colonial Agriculture in New York

Europe's leading seafaring countries—France, Spain, England and the Netherlands—all had their eyes on portions of the "new world" with hopes of bringing wealth from trade with the Native Americans and income with the establishment of colonies. Henry Hudson's 1609 exploratory voyage up the Hudson River gave the Dutch a strong claim to the surrounding area. Other Dutch ships followed up "The River of Mountains" in 1610 and 1611, but by 1612 there were only a handful of huts that had been constructed on Manhattan, crude homes for the few Dutch that had stayed to trade with the natives. A number of vessels sailed from Holland in the succeeding decade, bringing supplies and returning with furs, but the Dutch efforts to colonize the area were basically non-existent.[1]

As a means to increase trade and spur settlement, the Dutch East India Company was formed in 1621. The company's main purpose was commercial development of the colony, called New Netherlands, but its charter expressly provided that it was "to advance the peopling of the fruitful and unsettled parts."[2]

In 1624, the Dutch East India Company's first group of Dutch settlers came to present day New York State. Several families settled in our current New York City area and 18 families settled at Fort Orange, now Albany. A year later, three vessels from Amsterdam arrived carrying horses, cattle, swine, sheep, seeds, plows and other farming implements.[3]

During their first years in New York the Dutch maintained good relations with the natives, from whom they purchased food until their own settlers established agricultural production. The Dutch East India Company did not anticipate it would be necessary to send supplies to the colony after 1626, but the colony became self-sustaining very slowly and remained an economic disappointment to the company throughout the years of Dutch control. Population growth of New Netherlands was also slow, reaching only 270 in 1628, 300 in 1630, 500 in 1640 and less than 9,000 by the time of the English conquest in 1664. By contrast, in 1664 the population figures for Virginia and New England were about 40,000 and 50,000 respectively.[4]

New Netherlands settlement was limited because of the minimal opportunities and rewards for settlers, coupled with excessive hardships and danger from the natives. The land policies of the Dutch also inhibited the growth of the Dutch colony.

Agricultural advances occurred slowly in the colony. The Dutch introduced buckwheat to the colony in 1625. The main reason for its introduction was to make buckwheat cakes, but it was also used as a feed for poultry and farm animals. Wild flax had been found in Virginia by the first colonists and was soon introduced in New Netherlands. Flax fiber was shipped from New Netherlands to Holland as early as 1626.[5] By 1626, grain was evidently being processed into flour, as there is reference to a

1   Roberts, Ellis H. *American Commonwealths, New York* p. 24-26
2   Ibid. p. 32
3   Ibid. p. 36
4   Kammen, Michael *Colonial New York* p. 38
5   Carver, T. N. *Historical Sketch of American Agriculture in Bailey's Cyclopedia* p. 70

church congregation meeting being held in the upper floor of a horse mill. New windmills and a brewery were constructed in the 1630s.

In 1629, to increase agriculture in the colony, the West India Company put forth a charter of privileges and exemptions. It provided that any member of the Company might secure any unoccupied land in the colony, extending 16 miles along the coast of a navigable river on one side or eight miles on both sides of the river as far into the country as the situation of the occupier would permit. The charter also stated that the occupier must purchase the land from the natives and establish a colony of 50 persons on the property within four years. The founder of such a colony was called a "Patroon" and the colonists brought in were treated like the subjects of the "Lord of the Manor" in the old feudal systems in Europe. A number of such colonies were established along the Hudson River but did not survive. The exception was Rensselaerwyck Manor, extending on both sides of the Hudson with Fort Orange, now Albany, in the center. Killien

Van Rensselaer of Amsterdam, Holland, a director of the West India Company, was the purchaser but never came to the New World. He employed agents to trade with the Indians, giving goods in exchange for nearly 700 square miles of land. It was the largest of the patroonships and the only one to survive after the English assumed ownership of New Netherlands, renamed New York, in 1664.[6] This land purchase later became a large part of our present Albany, Rensselaer and Columbia counties.

After Van Rensselaer purchased the land in 1630, he sent colonists to settle on the property and provided them with cattle and farm implements. Within a few years there began to be some export of agricultural products, including tobacco, from the colony. Emphasis had been placed upon the fur trade prior to this time, since the fur trade provided greater returns.

Information from the Greenwich Village Society for Historic Preservation indicates that in 1629 Wouter

---

6   Hedrick, Ulysses Prentiss *A History of Agriculture in the State of New York* p. 42-43

An early view of Albany, courtesy of the New York State Archives

Van Twiller received a grant from the West India Company for a tobacco plantation at Greenwich and that the tobacco grew well. In the historic records of Harlem, a March 4, 1658 ordinance read, "the new village (Harlem) be formed for the further promotion of agriculture and cattle pasturing thereon".

In 1626, eleven slaves had been brought to the colony and in 1647, the West India Company recommended that slave labor be encouraged. By 1664, when the English took control, there may have been as many as 700 black slaves out of a total population of less than 9,000. The Dutch had found it expensive to bring white agricultural labor to the colony. In addition, sooner or later most whites applied themselves to a trade and neglected agriculture. A census of the total population taken in 1698 showed 18,067, with a quarter of the people living in New York City.[7]

In 1640, a group of colonists from Massachusetts founded Southampton on the Eastern end of Long Island. This was the first English settlement in New York. Within a few years additional settlers came to Long Island from Massachusetts founding the settlements of East Hampton and Southold. Before long, livestock belonging to the East Hampton villagers populated the broader section of the Montauk peninsula. In addition to cattle, both goats and sheep grazed on Montauk but did not multiply because of the large number of wolves and wildcats. Swine seemed to be able to avoid the wild animals and survived, roaming freely through the forests.[8] Huntington, Long Island derives its name from a hunting reservation granted to the Indians in an early purchase from them.[9]

These early farmers, isolated on the eastern end of Long Island, needed to be self-sufficient and capable of supplying their families with both food and clothing. They, like the first settlers a century or two later in Western and Northern New York, had to provide most of their own needs. The ox supplemented the muscle of the settler, furnishing the power to plow the fields and pull the loads on the farms since horses did not become of significance importance until later.

The raising of cattle was an important agricultural enterprise on Long Island from the time of the first settlements. In 1643, John Carman and Robert Fordham bought 60,000 acres from the Native Americans. The area known as the Hempstead Plains and now the site of Garden City and other communities, was utilized for the pasturing of cattle, sheep and hogs, which were driven from there to New York City for sale.

In the latter 1600s, the East Hampton town fathers purchased land east of the Town, totaling 11,000 acres, extending from Napeague to Montauk Point. The land was held in common by the citizens who would pay to have their cattle and sheep driven out to the pasturelands, where they would stay from spring to fall and then be driven back to their owners. This practice continued until 1879, according to an unpublished history written by William B. Jackson in 1942. Three men were chosen to stay with the animals while the livestock were at the pasture and three houses were built in different locations for the men's use. Earmarks were used to identify the animals, a common practice in the eastern United States, rather than the use of brands.[10] A comment made in the *1820 Gazetteer of New York,* concerning Suffolk County, was that it was peculiar that livestock pastured on the east end of Long Island have no flies to annoy them.

Comments from the diary of Ebenezer Miller between 1762 and 1768 offer a general picture of a farmer's work on Long Island at that time.

> *In the spring he mended fences and plowed his fields to plant hay, flax, oats and corn. By midsummer he was getting in his first crop of hay, harvesting rye and wheat, and a little later pulling his flax, which would be spun to make linen. In September he made the first cider of the year, some of which would be drunk fresh, the rest barreled and which, when fermented, would be a potent brew. Late in September he gathered corn. As the weather got colder he gathered wood to supply him for the winter and gathered dung to fertilize the fields. In the winter it was time to prepare the flax for spinning, to thresh the oats and wheat, and to slaughter sheep, hogs and cattle for food. Wood was chopped and piled in the woodshed.[11]*

7   Kammen, Michael *Colonial New York* p. 58 and 145
8   Ibid. p. 38-39
9   Hedrick, Ulysses Prentiss *A History of Agriculture in the State of New York* p. 43

10   Newsday.com August 25, 2008 *From the New Netherland Institute* Albany
11   Ibid.

It was uncommon for women to keep a diary in the 18th century but Mary Cooper was an exception. Her diary, now in the Oyster Bay Historical Society, provides a picture of her life on Long Island during the period from 1769 to 1772. During this time period much of what was used in the home was produced on the farm. The daily work started at sunrise and continued past sundown. Cooper writes of

> *...cleaning house, cooking meals, drying apples, making sweetmeats and sausage, salting beef, washing and ironing clothes, baking mince pies, taking care of honeybees, making wine, drying cherries, processing flax, making soap, sewing, picking blackberries, making candles and boiling souse, or pickled meat.*[12]

The Earl of Bellemont, governor of New York in 1701, wrote that,

> *"Not less than seven million acres have been granted away in 13 grants, and all of them uninhabited except Mr. Van Rensselaer's grant which is 24 miles square and on which the town of Albany stands."*[13]

These large grants extended from Yonkers northward on both sides of the Hudson River. It was anticipated that the grant owners would attract settlers to their property but few settlers could be induced to come and many of those that did left because of excessive payments to the owners and lack of opportunity to own their own property.

When the British took control from the Dutch, Fort Orange became Albany and Fort Amsterdam became New York City. In 1686, the principal towns in New York were New York City, Kingston and Albany, maintained wholly by trade with the Indians, England and the West Indies. England took furs, oil (flaxseed) and tobacco, while flour, bread, peas, pork and sometimes horses were shipped to the West Indies.[14]

In 1708, it was estimated that three-fourths of the linen and wool clothes worn by the settlers were made in their homes. England wanted the settlers to purchase all of their clothes from English manufacturers, but most of the settlers were too poor to afford English clothing.[15]

The land policies of the English had a tremendous effect in limiting the colonization of New York. They required title to land be purchased from the Indians. The English also expected the purchaser to obtain permission from the governor and the colony's council in order to buy land. If permission was granted a treaty was made, an Indian deed was obtained and a survey was required by the surveyor-general. There is little information available as to purchase prices of the land in these colonial times, but it is likely that permission was given by the governor to purchase land with some favor anticipated in return. The payments received by the Indians were always small. Hedrick provides a table of land patents, from 10,000 to 100,000 acres, choosing not to list the many patents of lesser acreage, granted by the English in Eastern New York between 1722 and 1775. There were 82 patents in this list amounting to a total of over two million acres. These patents covered large portions of eastern New York. The English patents and the huge land grants by the Dutch, coupled with the Indian threat west of Albany, left a limited amount of desirable land for the settler who wanted his own land. The Colonial Governors generally did little to encourage interest in agriculture. They very often entered office poor and left the office of governor as wealthy men, indicative of where their interests lay.

Governor Hunter was an exception to this policy. In 1711, a major colonization event occurred when he induced several thousand settlers to come to New York from districts in Germany called the Palatinate. These people, the Palatines, had been ravaged and impoverished by French invasion and came to New York as laborers. The larger part of the group was taken to lands along the Hudson River, about 100 miles north of New York City, while 357 remained in the city. The ones taken up the Hudson were to produce tar and turpentine and be furnished with bread and beer. Dissatisfaction in the arrangement was soon evident with many complaints from the settlers. In 1722, the Indians offered those Palatines land in Schoharie at easy terms and many moved to that area. Another group of Palatines came to New York in 1722. Many of them settled in the Mohawk Valley, where Palatine Bridge, the town of German Flats and the original patent of Stone Arabia preserve their memory. As late as 1861, *The American Agriculturalist* was printing its papers in both English and German for the benefit of these

---

12   Ibid.
13   Hedrick, Ulysses P. *A History of Agriculture in the State of New York* p. 44 and 49
14   Roberts, Ellis H. *American Commonwealths New York* p. 195
15   Ibid. p. 234

German settlers as well as those in Pennsylvania. The German settlers became excellent farmers adding much to New York State's prosperity. [16]

It is difficult to find in-depth statistical information concerning agriculture in colonial New York State. It was limited geographically to Long Island, the areas close to Manhattan, on both sides of the Hudson River, and later in the early 18th century, along the eastern Mohawk Valley. The Dutch had focused on the fur trade and profits. The English followed a similar pattern until the appointment of Colonel Thomas Dongan as Governor in 1682. Soon after his appointment many large tracts of land were patented and settlers began to be brought in, most notably the Palatines, but also some French Huguenots, English, Irish and Scots. These settlers, coming from different backgrounds, did not always look kindly on each other and conflict was common between neighbors in mixed ethnic communities.

In colonial times there were no agricultural censuses and few records of the amount of items being imported and exported. Agriculture is often overlooked in historical accounts of life. People have always eaten, resulting in agriculture being regarded as an everyday common event, not important enough to be recorded. When records are kept of the kind and quantity of imports and exports, we obtain a better idea of what agricultural products were being produced for market. With a census taken by New York in 1825 we begin to obtain a clearer view of what was happening on the farm.

Kamman provides figures of the number of boats and tonnage leaving the port of New York. He

Grist windmills at East Hampton, Long Island. A windmill provided the power to grind grain when water power from a stream was unavailable.

indicates that each year, between 1714 and 1717, there were an average of 64 vessels totaling 4,330 tons of goods. By 1721, there were 215 vessels carrying 7,464 tons. Kammen provides a comparison of the value of imports and exports between New York and England. The value of exports from New York remained fairly constant from 1715 to 1740, while the value of imports more than doubled. Between 1715 and 1718, New York's export tonnage with the British continental colonies equaled exports to Britain. Exports to the West Indies were double of those made to Britain. In 1707, whaling became a major industry on eastern Long Island with 4,000

16 Roberts, Ellis H. *American Commonwealths New York* p. 235-237

wooden barrels produced to store the whale oil produced from captured whales.

There is a 1723 quote in Kamman's book made by Cadwallader Colden, surveyor general of the New York Province, to London officials:

> *"The Staple Commodity of the Province is Flower &*
> *Bread which is sent to all Parts of the West Indies we*
> *are allowed to trade with, besides Wheat, Pipe Staves*
> *& a little Bees Wax to Madeira, We send likewise*
> *a considerable quantity of Pork, Bacon, Hogshead*
> *Staves, some Beef, Butter & a few candles to the West*
> *Indies."*

Later in his quote, he mentions the rich soil for producing hemp and that there are areas in New York where the soil is so rich that fine crops can be produced on the same soil for 50 years without dunging (applying manure).

Wampum was commonly used in trade with the Indians but by 1701 it ceased to be used as a medium in the colony. In 1709, the provincial government issued paper currency for the first time and this added fluidity to the commercial life of New York. [17]

Caleb Heathcote, a wealthy resident of Westchester County, constructed gristmills, a leather working shop, a fulling mill, a linseed oil mill and a sawmill in the early 1700s. He also raised flax and hemp and speculated in real estate. Adolph Philipse inherited a sizable fortune from his father who had married a rich widow. When his father died in 1702 he inherited substantial assets including 90,000 acres of fertile Hudson Valley land. Adolph lived until 1745 and during those years the number of tenants on his lands grew from 200 to 1,100. He also had 23 slaves. The tenants and slaves brought barley wheat and corn to his mills along the Pocantico River. He added bolting equipment and a coopering shop for the production of barrel staves. Much of his flour went to his bakery for the production of ship biscuits. He also added a dairy room to his manor house. A few merchants and land speculators like these prospered greatly, but the majority of settlers were fortunate to just eek out a living. [18]

The quantity and value of agricultural exports from New York for the period between 1765 and 1775 are listed as follows: 250,000 barrels of flour and biscuit for 250,000 pounds; 70,000 quarters (one fourth of a barrel) of wheat for 70,000 pounds; beans, peas, oats, Indian corn and all other grains totaling 40,000 pounds; salt beef and pork, hams bacon & venison totaling 18,000 pounds, 30,000 pounds of beeswax for 1,500 pounds; tongues, butter & cheese totaling 8,000 pounds, 7,000 hogsheads of flaxseed for 14,000 pounds; horses and livestock totaling 17,000 pounds. The total value of the exports for the 10-year period was approximately 418,500 pounds sterling. [19] The agricultural exports from Pennsylvania, during the same period of time, were about double those of New York while those from the New England colonies were relatively insignificant.

New York was not looked upon favorably as a place to settle by the Europeans and potential settlers from the other colonies for several reasons. The English deported many criminals to work as laborers in New York. In addition, New York was known to be a favorite spot for pirates. Thousands of settlers who did come were indentured for a number of years to pay for debts, which often included payment for their passage from Europe. A slave market was maintained on Wall Street and in 1723 more than 6,000 slaves, not all of them black, were sold. The slave trade would likely have continued if not for the unfriendly climate and the diversified crops grown in New York limiting the demand for slaves. It is no wonder that New York received a bad name in Europe and throughout the colonies!

A description by Hedrick of New York's colonial agriculture does not improve the image of colonial life.

> *"Livestock was poor of breeding and illy kept. The*
> *bull-plow was a crude wooden instrument reinforced*
> *with iron. The hoe was a commoner tool than the*
> *plow; the hoe-blade was made by the smithy, heavy ill-*
> *formed and clumsy, the handle a stick cut from the*
> *forest with the bark left on. The cradle was not in use*
> *until after the Revolution and grain was cut with a*
> *sickle; grass with a scythe. At least 80 percent of the*
> *inhabitants of New York before the Revolution were*
> *farmers, in the sense of living on the farm, but farmers*
> *were turning their hands to making something or*
> *doing something in a dozen trades. Parasites of the*
> *government, younger sons of the English nobility, and*
> *the Indians were the only gentlemen of leisure. Wives*
> *and daughters took an important part in managing*

17 Roberts, Ellis H. *American Commonwealths New York* p. 174
18 Kammen, Michael *Colonial New York* p. 173-174

19 Carver, T. N. *Historical Sketch of American Agriculture in Bailey's Cyclopedia* p. 49

the machinery of farm life. The good American traits of industry and perseverance were germinating in the work of subduing the forests and making a livelihood in the trades and on farms. "[20]

Hedrick also lists additional information that provides a clear snapshot of what colonial agriculture was like. From the beginning most colonial settlers raised corn. It was fed to horses and hogs and ground for meal. Almost every farm hollowed the stump of a hardwood tree to serve as the mortar and fastened a long cylindrical pole to a branch of an overhanging tree to serve as the pestle. The spring of the tree branch made the pounding of the corn into meal much easier. Wheat was the staple crop with 10 bushels an acre considered a good yield. Rye and barley were grown but were of lesser importance.

Cattle were left to graze in the forests and as a result were in poor condition and gave little milk, perhaps a quart or two a day. Because their diet consisted of a variety of wild shoots and grasses the butter made from their milk was of poor quality. There were a mixed variety of cattle with the Dutch bringing some from the island of Texel and a mixture of English, Danish and Spanish breeds. English farmers visiting thought the lack of grass was the greatest handicap to American farming. Sheep were not nearly as important as in the 19th century because of the lack of grass and the danger from wolves and wild dogs. Generally the sheep of the colonists were very poor, yielding little wool and looking as much like goats as sheep. If wolves and bear were not excessively plentiful the colonists' hogs ran wild in the woods.

Apples were grown in abundance for both eating and producing cider. Large quantities of cider were made to last a family a whole year. Pears were commonly grown and peaches too, where the climate permitted. Wherever peaches were grown, peach brandy was made. The settlers grew cherries and grapes but the American species of grapes and berries were not domesticated until about 1800. It wasn't until after 1800 that potatoes and beans were grown in any quantity.

Since land was cheap the settlers had little interest in maintaining its fertility. The use of manure for fertilization was largely ignored and there was limited crop rotation. Most of the crops and animals were grown for the family's consumption or to barter for necessities. About the only source of cash was for forest products.

In 1768, Thomas Young established a commercial nursery at Oyster Bay on Long Island and about the same time, William Prince established a nursery at Flushing on Long Island called Linnaean Botanic Garden. In 1771, the latter published a list of the varieties of fruit trees for sale including 24 apples, 9 apricots, 18 cherries, 12 nectarines, 29 peaches, 42 pears and 33 plums. Five of the apples originated in this country but all of the rest of the fruits originated in Europe. Tree fruits of Europe do not thrive here so it is likely that fruit growing was not very successful in the colonial period. William Prince's nursery business prospered, however, and for more than 100 years was the leading nursery in New York and perhaps in the United States.[21]

The ox, which was the chief draft animal throughout the colonial period, was used for plowing, harrowing and pulling a loaded sled or cart. The ox was a multi-purpose animal including serving as food for the family after completing its days as a beast of burden. The ox, like the cow, its female counterpart, was allowed to graze in the forest on whatever nature provided costing little to maintain.

The settler's family worked as a unit, all occupied with the task of survival and the hope for a better life. A large family was an asset and children were put to work at an early age. Ben Franklin, writing before the Revolution, said that a widow with a half a score of children was an object for the fortune hunters of America.

The farm families of Colonial New York made great sacrifices in the War for Independence. Many left their farms to fight in the war, never to return. The women left behind, not only took care of their normal responsibilities, but stepped up and carried the burden normally shouldered by the men. The women, children, and remaining men on the farms endured the dangers of surprise attacks, while producing food and fiber for their families and the needs of the Colonial army. Salt pork, wheat, linen, wool, and leather were as necessary to the conduct of the War as guns and powder.

20 Hedrick, Ulysses Prentiss *A History of Agriculture in the State of New York* p. 64

21 Carver, T. N. *Historical Sketch of American Agriculture in Bailey's Cyclopedia* p. 72-73

The Colonial government had little silver or gold so issued paper currency, which as the war waged on steadily decreased in value. It took self-sacrifice and supreme faith in the American cause, for the farmers to exchange their products for the almost worthless paper promises of the Colonial government.[22]

Records show that the Army, in 1776, collected in New York City, 106 barrels of pork, 97 barrels of beef and 1,694 barrels of flour. By 1779, the Army often had to impress cattle, flour, grain, oats, hay, cows and beef from farmers and dealers to obtain these products. It is estimated that New Yorkers lost approximately $5 million by selling produce for paper money. In 1777, to obtain the fiber to make rope, hemp seed had been given to farmers for planting.[23]

Our life is so different today that a few more of the facts of country life in colonial times need to be mentioned: there was almost no travel, other than by water, and roads were almost non-existent except in the cities and villages; Native American trails were the common roads, supplemented with paths worn traveling to a neighbor's farm or to a nearby settlement; the family's clothing was coarse homespun made from flax, hemp or occasionally wool; meat needed to be salted if it was to be stored for any length of time and fresh meat was an unusual but enjoyable treat; honey from beehives and maple sugar were the only sweeteners; and barter was the chief form of trade. Mail service was almost non-existent and settlers seldom if ever heard from the families they left behind.

Today, we regularly rely on the goods and services produced by others to fill our needs and pleasures, which is in total contrast to the rural settlers of Colonial times who of necessity were self-sufficient. Those settlers produced most of their needs as a normal part of their daily lives.

---

22 New York Division of Archives & History *The American Revolution in New York* p. 179
23 Ibid. p. 179,185,197

Chapter III
Wilderness to Farmland

New York declared itself an independent state on July 9, 1776, near the beginning of the war for independence from England. For the purposes of this book, however, the end of the Colonial Period occurs when New York becomes the eleventh state in the Union with its ratification of the United States Constitution on July 26, 1788. The Colonial Period had lasted 164 years and with the war now past, New York was destined to become the most rapidly growing state in the Union, reaching dominance worthy of it later being called the Empire State.

New York is a relatively large state when compared to the states in New England and some of the Middle Atlantic States. It consists of 49,204 square miles, extending 326 miles from east to west, and 300 miles from north to south. In addition to many miles of sea surrounding Long Island and greater New York City, it has 75 miles of border along Lake Erie and 200 miles of border along Lake Ontario. These bodies of water have often determined what crops will be grown in their vicinity. A relatively small part of New York's land had been utilized for agriculture prior to the Revolutionary War but following the War, vast areas were opening for agriculture.

During the Colonial Period, New York could be compared to a genie in a bottle. With French settlements on New York's northern border and France a bitter enemy of England, settlement in upstate New York was not prudent. In addition the Iroquois, composed of several powerful Native American tribes, was the entity of power in Upstate New York making settlement very dangerous. At times the natives aligned themselves with the French, and at other times with the English, but seldom were they on the side of the settlers who moved onto their lands.

In 1763, after the English had defeated France in the French and Indian War, England forbid the settlement by colonials west of the mountains. The English feared that the Iroquois, continually stirred up by the French, would maintain constant warfare against the settlers. In addition, throughout the Colonial Period, the land policies of both the Dutch and the English had limited the opportunity for common people to settle in the State. Both the Dutch and the English had sold, for money or favors, large land parcels to the wealthy, who often didn't even live in the colony. It was common practice that the settlers on these tracts could only rent land and were forever beholden to the owners. After General Washington, in 1779, sent several thousand troops, commanded by General Sullivan, to destroy the Iroquois, coupled with the English army surrender at Yorktown in 1781, upstate New York became ripe for settlement. Many of the soldiers with Sullivan took note of this rich land and came back to settle some years later.

Settlement of the vast lands of Upstate New York didn't happen immediately. The new state had just been through a long, expensive war and there were many pressing needs to consider. New York now owned great quantities of land. Its leaders, however, looked upon this valuable resource in terms of the money they could receive from its sale. In 1791, a law was passed authorizing the state commissioners of the land office to sell any of the public lands at their discretion. This resulted in the sale of over five million acres for about one million dollars. Placing this great quantity of land on the market at one time provided great bargains for buyers but brought relatively little money to the State. A prime example is the purchase, by Alexander McComb, of over three million acres of this land, most of which he purchased for only for 8 pence an acre.[1] (A pence was worth a little over one cent.)

Massachusetts, at a much earlier date, had claimed all the land from the Atlantic coast to the Pacific coast. In a 1786 agreement, it ceded to New York sovereignty of six million acres west of a line through Seneca Lake to Sodus Bay in turn for the ownership right to sell the land. In 1787, Massachusetts sold this land to Phelps and Gorham for one million dollars subject to approval by the Seneca natives. Phelps and Gorham were able to obtain title to only the eastern one-third with the remainder reverting to Massachusetts, which sold it to Robert Morrison in 1791. In 1793, Morrison sold the western 3½ million acres to the Holland Land Company and reserved 500,000 acres, which later was sold to Sir William Pulteney. A strip 2½ miles wide and more than 200 miles long bordering Pennsylvania known as "The Gore", was purchased by a Connecticut Company. Massachusetts also received and then sold 230,000 acres between Oswego and Chenango.

The sale of over 11 million acres of New York land, in just a few years, by Massachusetts and New York represented over 25% of New York's land! Wealthy individuals or organized groups purchased this land in huge blocks with just one purpose in mind, to make money. To accomplish this they needed buyers and there were great numbers of people who wanted the opportunity to own land.

Another 1¾ million acres of State land was granted to Revolutionary veterans. This land, known as "The Military Tract", consisted of what are now the counties of Onondaga, Cortland, Cayuga, Seneca and portions of Oswego, Schuyler, Tompkins, Yates and Wayne. This land was divided into 28 Townships, each with 100 lots of approximately 600 acres. Several lots in each Township were reserved for the gospel, schools and to provide for land covered by water. The veteran who did not pay for the survey of his lot was required to provide 50 acres of his land in payment.

In 1776, the Congress made provision for 88 regiments to carry on the Revolutionary War with a quota of four regiments to be provided from New York. By 1781, only two New York regiments had been raised and to encourage enlistments the New York legislature voted to raise the other two regiments by offering land as an incentive. Since the State had little hard currency but great quantities of land, they agreed to give 500 acres to each man who enlisted for three years. Officers were to be given greater amounts depending upon their rank. The War officially ended with the Treaty of Paris in 1783, but because of difficult negotiations with the Onondaga and Cayuga tribes it took until 1789 before the ownership issues were resolved. It took another year to survey the lots and finally the lands were assigned to the veterans in 1790. Congress earlier had passed a resolution granting 100 acres in Ohio territory to all soldiers remaining in the army until the end of the war or until discharged by Congress. The New York soldiers who did not take their 100 acres in Ohio were entitled to another 100 acres in New York's Military Tract making a total of 600 acres available for each man.

Only a small percentage of the Military Tract recipients settled upon their land. Prior to 1790, the veterans had received script (an "I Owe You") for land promised to them. The war had ended seven years prior to the allocation of the land and the Military Tract was just a wilderness without roads. The soldiers, disillusioned many times by unfulfilled promises of both Congress and the State and feeling that the script was almost worthless, often traded it for something of little value.

The Military Grants brought a large influx of population into Central New York. The veterans who did settle upon their grant usually sold off parcels to other settlers, as 600 acres was far larger than could be utilized by one family. The speculators that bought the script from veterans also sold off portions to others with the result that land was now available for thousands of families to start a new life in an undeveloped wilderness.

In addition to the veterans of the War, there were English loyalists looking for a new beginning as well as English soldiers and their German allies who saw

---

1   Roberts, Ellis H. *American Commonwealths, New York* p. 451

Albany Nursery, 1826 Broadside listing dozens of fruit tree varieties for sale. J. Buel, one of the proprietors was the publisher of The Cultivator, *for the New York State Agricultural Society. Courtesy of Albany Institute of History & Art, PB0051.*

opportunity in the new state. The better land in New England had been settled and the rich lands of Central and Western New York beckoned settlers. Word had spread in Europe about the vast quantities of rich land available, and immigrants poured in from northern Europe. Between 1783 and 1790 the population of New York increased from 233,896 to 340,120, almost 50%! By 1800, it increased to 589,051, more than double its population only 17 years earlier. In 1810, the population was close to a million, exceeded only by Virginia, which New York surpassed a few years later. New York State was growing more rapidly than any other state.[2]

A road from Whitestown to Geneva was opened in 1796 because of the large number of settlers moving westward. In 1798, a road was opened from Genesee to Buffalo and to Lewiston. These roads were nothing like the ones we are familiar with today. Trees had been cut and stumps were removed to allow sleds pulled by oxen and the occasional team of horses pulling a wagon or sled to pass through.

After the Revolution, Albany, the sixth most populated city in the US, with a population of 3,050, became a center of trade. Its location near the Mohawk and Hudson Rivers made it both a natural terminus and shipping point. During the War there had been no exports from New York as the English controlled the New York City harbor, but by 1791 exports totaled $2½ million, increasing to $14 million by 1800 and to over $17 million by 1820. Of the $17 million from trade in 1820, $10 million of those exports were credited with coming from the soil![3]

While farmers were busy growing crops and producing livestock, there was much activity inside the homes of New York residents. The State estimated the value of manufactured goods in 1811 at $30 million with household production amounting to $12 million of that figure. In **Table A,** in the Appendix, you will note the yards of cloth produced in the homes of each of the counties. In 1810, over eight million yards were produced in the homes with an estimated value of close to five million dollars.

The large tracts of land that were rapidly broken down into smaller units, coupled with the rapid influx of settlers, presented many problems for the State, local municipalities and even the landowner. Often an early settler occupied a parcel of land, in what was a wilderness, with the impression that it was his since no one occupied it. Later when the rightful owner appeared the squatter would have to be evicted. Unscrupulous landowners often sold the same property, located miles away in the wilderness, to two or more buyers. In addition, with millions of acres being surveyed by a multitude of surveyors not all well trained, there were errors that had to be settled in the courts. Often a tree or stump marking a corner on the original survey was later removed making subsequent boundary location difficult. For many years court dockets were filled as attempts were made to determine rightful landowners.

The owners of large tracts of land, dating back to the Dutch and continuing under the English and even into the 1800s after New York became a state, offered settlers land, but with strings attached. The agreements were generally more favorable to the landowner. A variety of stipulations, in addition to rental payments, were required, often without the tenant's realization that he could never obtain full ownership to the property. Government did not discourage these policies, as a portion of the rents went to the colonial government and later as a source of revenue for the State. When rents were not paid, the local municipalities forced collection or had the tenants removed from the property. In 1839, organized protests to civil processes of collecting rents forced the governor to send State troops to assist civil authorities in rent collection. These organized protests originated in the older tracts in Eastern New York but later became common in some of the tracts in Western New York. It was not until 1850 that the State removed the last of these onerous landlord rights. Land disputes between the purchasers of the large tracts and the settlers resulted in many years of disputed land ownership court cases.

Since forestry is a form of agriculture and almost the entire State was initially covered with trees, trees could be considered our first agricultural crop. Many of the trees were giants by today's standards. There were white oaks up to five feet in diameter. The large trees had never been cut and continued to grow until old age or disease brought them to the ground. The Native Americans had little use for the huge trees, preferring trees less than a foot

---

2   Roberts, Ellis H. *American Commonwealths, New York* p. 453
3   Ibid. p. 455

Engraving of an early farm. Note the well sweep on the right and the rail fence surrounding the home rather than the cattle. From W.W. Clayton's book, The History of Onondaga County 1615-1878.

in diameter, which were much easier to cut and handle, for their stockades, fuel and other needs.

Most white settlers considered the abundance of trees an obstacle that needed to be removed. Food was needed for their families and eventually they hoped to produce a surplus of food for sale. In the parts of the State where the land could be tilled, the settlers removed the majority of the trees, leaving only enough to take care of their future lumber and fuel needs. Seldom would a farmer clear-cut his land. Even on the smaller farms of less than 100 acres, five or ten acres were left wooded to provide for future lumber and fuel needs. If part of the farm were hilly or swampy and difficult to till, that area would be left for the woodlot.

The main uses for the wood of the forest were; logs for constructing buildings, ashes for making potash, fuel and lumber. Secondary uses included tanners bark, fuel for the production of charcoal and ashes for improved soil fertilization.

The first use of the forest by the settlers in an unpopulated area was the construction of a log home. The only materials available to the settlers were trees and the earth upon which they grew. Both were used to make the first buildings. Logs from 7 to 12 inches in diameter were cut for the sides. Logs of a similar size were split to form the roof. Bark was peeled from basswood, elm or ash trees to provide the covering for the roof. Placing logs of this size into place was a difficult task without neighbors to help. They had to be rolled by man up a ramp into position unless there was an ox or horse to furnish the power. Other buildings were constructed in the same manner until a sawmill was constructed.

Living on the dirt floor provided by nature was common until the settler had time to split logs and smooth them with an adz, forming a puncheon floor, which was dryer and cleaner. Clay or moss was used to fill any cracks between the logs. Sometimes, when shelter was not needed immediately, a broad axe was used to square off the sides of the logs. This made a neater looking log building and required only minimal chinking between the logs. The first settlers in an area had no glass so instead used greased paper to cover window openings to allow the entrance of some light. The cross cut saw did not come into general use until the latter part of the 1800s so all the cutting was done with axes.

The typical log cabin was 16 by 24 feet. The cabins of the first settlers in a community were normally one story because this size cabin could be constructed by one family and would provide shelter fairly quickly. It was not uncommon for a family to make a temporary shelter from tree boughs until the cabin was constructed. There was no need for a barn until livestock numbers increased and there were crops to store. If a barn was needed before there was a sawmill in the area, the barn was small and made from logs. Often a barn was the first frame structure on the farm, since it was constructed later than the log cabin. The New York census of 1855 lists 33,092 log homes with an average value of about $40. In 1875, there were 12,659 log homes remaining with an average value of $82.

Another important use for logs was for water pipes. A long auger was used to bore a hole through the center of the log and the end of one log was hewn to provide a tight fit into the next log. Log pipes were used to transport water from springs to homes and to livestock. Log pipes were even used in the early villages and cities that had public water supplies.

When there was not a market for logs, which was often the case during the early settlement years of a community, a large area of the forest was cut at one time for burning. A common practice was to fell the trees in a manner to make the tops fall onto each other in a center of a circle. Then, after the tops had dried, the trees were set on fire with most of the tops consumed in the flames. Later the remaining trunks of the trees were cut into manageable lengths and oxen used to pull the logs together into piles for burning.

Hardwood ashes were a valuable source of potash and were often sold by the farmer to an ashery for six to eight cents a bushel. It has been reported that some farmers paid for their farms from the sale of the ashes from trees they burned to clear their land. The farmer transported his ashes to the nearest ashery where they were processed by leaching the ashes with water. The resulting lye was boiled in large kettles. After the liquid had evaporated, there was a solid mass remaining in the bottom of the kettle called "black salts." This material was dark in color because of carbon and other impurities. Sometimes the farmer did his own leaching and boiling, and then sold the "black salts" to the ashery, which paid him about $3.00 per hundred pounds. The asheries had brick kilns in which the "black salts" were burned to remove the carbon and impurities, leaving a fused, bluish-white, much purer product known as "pearl ash."[4]

This was very valuable and could be transported to Albany, New York City and to Europe. Pearl ash was an important ingredient in glass, soap and china production. It was also used for the scouring of wool, critical to England's large textile industry. Without question ashes were a substantial source of income to the settler ridding his land of trees for the planting of annual crops.

The 1825 New York census lists 2,105 asheries in the State. Logs for sawmills only made a dent in the millions of trees that stood in the way of the settlers urge to clear land for farming. The end result was hundreds of thousands of acres were cut and burned producing ashes as the end product. I have not found information as to the value of the ashes produced prior to 1845, when the 738 New York asheries produced product valued at $909,194. It is probable that the value of the ashes produced each year prior to that date was similar or greater because the improved farmland acreage increase prior to 1845 was comparable to the years after. The housewife often traded the ashes from her fireplace with the country store proprietor for items she needed in the home. The country store proprietor then sold them to an ashery. As the acres of forest being cleared decreased, the number of asheries decreased to 68 in 1855 and to 54 in 1865.[5]

The New York census of 1825 provides an interesting "snapshot in time" of the settler's battle against the forest, showing 5,195 sawmills in the State. A sawmill was needed within a few miles of every farm because oxen or horses were the only means of dragging the logs to the sawmill and returning with the sawn lumber. Occasionally a farm located adjacent to a river floated the logs to the sawmill. Rivers provided an efficient means for transporting logs to the various sawmills. Logs were fastened together and floated down the river in large rafts.

---

4   van Wagenen, Jared Jr. *The Golden Age of Homespun* p. 166-167
5   Ibid. p. 166-167

The owner of the logs used an unheated branding iron to mark his initials or brand on the end of the log. The branding iron had a wooden handle and a heavy iron head, which was swung against the end of the log leaving the owners brand on each log. If a log broke loose from the raft it could be identified in the river as well as later, at the sawmill. The burgeoning population in upstate New York needed lumber for homes, barns, barrel staves, bridges and a multitude of other uses.

Where today we have containers made from a variety of plastics and metals, barrels of wood were produced well into the 20th century to hold almost every product that needed a container. There were barrels for grain, flour, salt, apples, and potatoes. Lumber was sawn and the coopers of New York produced thousands of barrels each year. In addition to many full time coopers, barrel making was a part time occupation for numerous farmers in the early days of our State's history. The production of barrels provided a source of additional income during the winter months when farm activities were not as pressing. Staves for barrels were made by hand until after the Civil War, providing work for untold numbers of both coopers and farmers.

An often-ignored use of lumber from the forests was for the many bridges that were needed to cross the rivers and streams, all of which were constructed of wood. A prime example is the long bridge over the northern end of Cayuga Lake that was opened for travel in 1800. This bridge was part of the Seneca Turnpike and was over one mile long. It was 22 feet wide to permit stagecoaches to meet on the bridge and was supported on trestles 22 feet apart. Another example is a wooden bridge built in 1806 over the Hudson River at Troy.

When steamboats began traveling up and down the Hudson River wood was the fuel that provided the power. Still later, when trains came into use, huge quantities of wood were burned to power their engines until around the time of the Civil War. At numerous locations along the railroad tracks travelers saw huge piles of wood for refueling the train engines.

Wood was the only fuel available for heating both homes and commercial buildings. A large fireplace was constructed in buildings to burn wood for both cooking and heating. The pine plains on Long

A log branding iron, circa 1840. Striking the end of the log with the branding iron, marked logs with the owner's brand to identify them at the sawmill.

Island supplied a large portion of the fuel consumed in New York City. The 1820 *New York Gazetteer* stated that the town of Brookhaven in Suffolk County sent about 100,000 cords of wood to New York City annually for use as fuel. During the summer most cooking was done outside over an open fire to keep the log cabin homes from become unbearably hot. After the first year in the wilderness, the family cut their wood several months before it was needed to allow time for it to dry sufficiently and provide cleaner burning.

Bark for tanning hides for leather was a by-product of lumbering from the very beginning of settlement. The hides of farm animals, especially cattle, were important for their use in making shoes, harnesses and a variety of other leather products. Four-foot strips of the bark of oak and hemlock trees were stripped from logs in the late spring and early summer to use for tanning. The 1810 census of American manufacturers showed 867 tanneries and the 1835 New York census shows 412 tanneries in the State. This number indicates large quantities of bark were needed for tanning. In 1847, Zodoc Pratt of Prattsville in Greene County was said to have the largest tannery in the country, tanning 60,000 sides of cattle hides a year. He cut 10 square miles of hemlock forest, which yielded 18 cords of hemlock bark per acre.[6]

Another important use for wood was for the production of charcoal. In 1768, veins of anthracite coal were opened in Pennsylvania. With only

_____

6   van Wagenen, Jared Jr. *The Golden Age of Homespun* p. 191-192

wagon transportation to New York available, little was used. Since coal was not readily available, charcoal was the fuel of choice for the blacksmith, the tinsmith and the iron industry. It took about 2 ½ tons of charcoal to produce a ton of pig iron. Charcoal was produced in large quantities until about 1860, when it was replaced by coal transported to New York on railroads.

A charcoal pit used from 25 to 30 full cords of hardwood. Four-foot lengths of wood were piled on top of each other to form the shape of the old-fashioned beehive. This was then covered with earth and sod, leaving just a few holes for lighting and an entrance for air. Once the fire was burning well the holes were covered and the pit was watched day and night for 6 weeks to make sure flames did not break through the earth. Each cord of wood produced about 30 bushels of charcoal that sold for about 14 cents a bushel.[7]

The 1825 New York census lists a little over 7 million acres of improved land, which is land cleared of trees. The 1835 census lists over 9 ½ million acres and the 1845 census lists over 11 ½ million acres of improved land. The speed with which the forests were being cut diminished after 1855. Small amounts of land were still being cleared for farming until 1930, but the period of the most rapid clearing of land was prior to 1855. It is unfortunate we do not have the number of acres of improved land prior to 1825, but we can make a safe assumption that a large percentage of the 7 million acres of improved land at that time was cleared between 1790 and 1825. The acreage of improved land increased by another 2 million acres between 1845 and 1855 but by only another million during each of the next two decades reaching 15,875,552 acres in 1875. Almost all of this improved land acreage had been forests a century earlier. It is hard to imagine the tremendous change in the New York landscape during that 100-year period! The total acreage of all the farmland, including its woodland, pasture and wetlands in the State today, is less than one-half of the improved farmland that existed in 1875.

The woodlands of New York have continually produced substantial income for farmers since the first white settlers arrived. The wood harvest diminished once trees were removed from our potentially tillable land. There is a vast acreage of forest that was never cleared and planted because of the difficulty in tilling. The forests are continually producing trees for harvest by lumber companies and professional loggers. Even today, many farmers in the State have a woodlot where fuel for the farm home is harvested and where logs are periodically sold to the sawmill.

The acreage of New York's woodlands has been increasing since 1910. Land acreage in farms grew steadily until about 1880 when it totaled 78 per cent of the land area. In 1910, both State and privately owned woodland accounted for 47.5 percent of the State's land area reaching 70 percent in the 1990s. It is quite likely that the acreage of forests in New York is even greater today than in the 1990s.[8]

In recent years, the production of alternative fuels has been encouraged by our federal government. Research into the production of ethanol from a variety of crops, including hybrid willows, is taking place. Only time will tell if, in the future, our New York farmers will be growing trees to help meet the increasing need for energy.

7   van Wagenen, Jared Jr. *The Golden Age of Homespun* p. 161-163

8   Stanton, Bernard F. & Bills, Nelson L. *The Return of Agricultural Lands to Forest* p. 12

Chapter IV
The Self-sustaining Community

From a historical perspective, the number of years that the early settlers in Upstate New York existed in a self-sustaining community was relatively short, whereas the length of time must have seemed much longer to those who lived the experience. For example, Ontario County, now composed of 12 counties, encompassed the entire area of Western New York in 1790. There were no villages between Utica and Fort Niagara, and less than 1,000 white people living west of what is now the city of Geneva. By 1795, the number of white settlers had increased to 12,000 and by 1810 there were almost 300,000! The British had retained Fort Niagara and Fort Oswego, disregarding the agreement for the forts to be abandoned as stated in the Treaty of Paris. These British encouraged the Iroquois to war against the settlers. Finally, in 1794, General Wayne was victorious over the Iroquois and a treaty, never broken by either party, was signed between the six nations and the United States at Canandaigua. Two years later the British turned over Fort Niagara and Fort Oswego to the United States. With the threat of both the Iroquois and English removed, settlers poured into Upstate New York.[1] As the number of settlers increased roads were built, blacksmiths set up shop, stores were constructed and life began to replicate life in the villages and on the farms they had left behind.

The first settlers of Upstate New York relived the experiences of many of their forbearers, who over a century earlier had settled on Long Island, along the Hudson River and in the eastern Mohawk Valley. These settlers were almost wholly dependent upon themselves for survival. They needed to take most of what they needed with them to survive in the wilderness. Traveling along a trail or through the forest, where there was no road, a family was limited as to what they could bring. It was often only what could be loaded on a sled and pulled by an ox or a two-wheeled cart that could be maneuvered through the forest. Two essentials were a gun, for both protection and hunting, along with an axe to cut trees. A kettle was needed for cooking, a few knives, forks and spoons, a hoe, a blanket, some food and a supply of seed for planting. Since land needed to be cleared before planting, it was desirable for settlers to arrive at their destination in late winter. At that time of year the ground was likely to be frozen and there might be snow to allow the sled to pull easier. Sometimes the men in the family would go to their new land in the wilderness during the summer for a few weeks to cut and girdle trees, construct a crude log cabin, plant winter wheat and then return, bringing the rest of their family the next spring.

The first settlers in all areas of New York had to make their way through the forests as best they could often using Native American foot trails that evolved into rough roads. Streams had to be crossed, fallen trees removed, boggy areas circumvented, and mountainous terrain avoided. Travel that might take two weeks could be accomplished in two days after roads were cut through the wilderness.

---

1   Hedrick, Ulysses Prentiss *A History of Agriculture in the State of New York* p. 94-96

The majority of the early settlers, were poor and had few possessions. They were filled with hope for a better life in their new surroundings, but certainly often questioned their wisdom in moving. In Onondaga County, the wife of one Revolutionary veteran who settled on his land grant, didn't see another white woman for seven years. If a settler could secure land, little money was needed for other purchases. A yoke of oxen, at the close of the Revolutionary War, cost from $60 to $90 and the necessary farm tools and supplies could be purchased for $50. These tools included an axe, an adz, a fro, a length of chain and a hoe. As an increased number of families settled in a community small industries were established. **Table C** in the Appendix shows the number of naileries, flaxseed oil mills and breweries in each of the counties in 1810. Often if a farmer did not own a yoke of oxen, arrangements might be made to borrow a neighbor's oxen in exchange for his labor. Neighbors were dependent upon each other and it was a common practice to exchange labor.[2]

The early settler, with hardwood trees to turn into potash and who was able to grow a crop of wheat successfully, could prosper and pay for his farm in a relatively short time. Good land yielded 15 bushels of wheat per acre and if not sold or traded locally might be sold in the Albany market for $2 a bushel. A settler was especially blessed if there were several

sons in the family to help clear the land, plant the crops and harvest the potash.

Occasionally a lucky settler unknowingly purchased land that had previously been cleared by the Native Americans. The vast majority, however, had to clear the land they settled on. If there was not time to fell trees prior to planting, the settler girdled the trees with an axe preventing the flow of sap and the growth of leaves. Seeds could then be planted around the trees allowing sunlight to grow a crop. After harvest the settler had more time to fell the trees and burn them. After cutting the trees it was common practice to farm around the stumps for a number of years to allow decay and provide easier removal of the stumps. Harvesting the grain grown around the tree stumps was not difficult as it was cut with a sickle in the same manner as a field without stumps.

There were several advantages of oxen over horses. Oxen were able to survive in the forest by eating leaves from young shoots as well as tender bark. A horse needed hay or grass that wasn't commonly available. Oxen were excellent for hauling logs and pulling stumps because of their slow steady pull compared to horses. Another advantage of the ox was when it was too old to work or if injured it could be butchered and eaten by the family. Families with a cow that had a male calf often castrated the calf and trained it at little cost since it was a by-product of owning a cow. Horses, of course, were faster

---

2   Hedrick, Ulysses Prentiss *A History of Agriculture in the State of New York* p. 110

Harrowing with oxen, circa 1900. Courtesy of Toby Shelley, Town of Otisco Historian.

but were also much more expensive and generally owned only by settlers of some financial means.

An ox could be obtained from any breed of cattle and was usually a mixture of a multitude of ancestors. Devon cattle were introduced into Massachusetts in 1800 and came to New York's Otsego County in 1807. The Devon was originally a solid light red and won recognition as excellent oxen. The early census data groups all cattle under the title of "neat cattle" so we do not have early numbers of oxen. The 1855 census, at a time when oxen were being replaced by horses, shows more than 144,000 working oxen.[3]

Prior to the construction of a gristmill in the community, the settlers used an axe and adz to hollow out the stump of a large hardwood tree into a bowl-like shape. This would serve as the mortar and a hardwood piece of wood, about 5 inches in diameter and 4 feet, long was fastened to an overhanging branch of a nearby tree to serve as a pestle. Grain was put in the stump of the tree and a person held the pestle in both hands to pound the grain into flour. It was a long, tiresome process but provided some coarse flour for cooking. A horsehair sieve was used to separate the flour from the coarser particles of grain. We know from historical accounts that this method worked, but we also know it was done only out of necessity. There are numerous stories in the histories of our counties of a person traveling 10 miles and more with a bushel of grain on his back to a gristmill to have it ground. The load returning home was lighter, as the miller kept a portion of the grain as payment.

The crops grown by the early settler were similar to the ones they had grown in the areas from which they had come. Wheat was the first major crop and was as good as money for trade. Oats, barley, corn and rye were also commonly planted. All of these could be used as food for humans and livestock, and for distilling. They also could be easily stored and transported. During their first years on the frontier, few of the settlers produced extra food for sale other than grains. The farm family consumed all of the vegetables grown because there were no nearby markets and no convenient means of transporting them to distant markets. A cow produced milk for butter and cheese but the quality of the end product was often poor. The cow's diet consisted of a variety of wild plants that contributed neither quantity nor quality to the milk. Fruit trees were planted when the farm was first settled but it took a few years before they provided the family with fruit. Apples and pears were grown in abundance as well as peaches and currents. A major use of apples was for cider and peaches were usually turned into peach brandy.

During the years of early settlement the crops were fenced to keep the farm animals from eating them and the cattle, hogs and sheep allowed to run loose. The cattle and hogs survived quite well, but wolves made it necessary to keep the sheep confined. In almost every community a bounty was placed on wolves because of the great number of sheep they destroyed. Bounties on wolves of $5 to $20 were not uncommon. When you consider that $10 was the equivalent of a month's wages for a man, it is evident that wolves were doing a great deal of damage to the sheep. It was not only the sheep that were killed, but also calves and young pigs. In a number of instances, when a community became settled, the farmers would organize a wolf drive in an attempt to eradicate the wolves. New York's last great wolf drive took place in Tioga County in 1828.[4]

Sheep were of critical importance to the settler as they produced the wool used for clothing, especially cold weather clothes for the settler. Sheep, until into the 19th century, were not of high quality and had little wool on the legs or belly. The fleece from these sheep weighed around three pounds but the fleece of the Merino sheep, introduced in the early 1800s, weighed six pounds and was of much better quality. There was a Merino craze in the early 1800s with bidding of up to $500 and even to $1,500 for one sheep. This was an unimaginable sum of money for that period and was a price that almost no one could afford. In 1807, Merino wool sold for $1 a pound and by 1814, as much as $4 a pound but by 1845, the price was badly depressed with Merino wool ranging between 18 and 60 cents a pound.[5]

Farming practices had seen little change for centuries during the time much of New York was settled. In 1814 and 1819, Jethro Wood of Cayuga County received patents on a cast iron plowshare. The plows used at that time were mostly wood with a piece of metal at the point to extend the life of the wood.

---

3   van Wagenen, Jared, Jr. *The Golden Age of Homespun* p. 39-40

4   Ibid. p. 70
5   Ibid. p. 257-259

Flail used to thresh small grains and a winnowing board to separate grain from chaff.

An early wooden roller at the Farmers Museum in Cooperstown.

Much of the land was prepared for planting with spade and hoe. A large tree branch was often dragged to level the ground and cover the grain scattered by hand when planting. Seed drills were not available until the middle of the 19th century and even then most farmers still planted seeds by hand.

Grain was harvested with a sickle, back to the time that the pyramids had been constructed in Egypt and was still harvested that way by our early New York farmers. A scythe was used to cut hay but was not satisfactory for grain as it scattered the stalks, making binding into bundles difficult. Some farmers used a sithe, which had a blade similar to a scythe but was held in one hand, while a grain hook was held in the other hand pulling the stalks together in the same direction. The estimated date for the introduction of the grain cradle is about 1776, with grain cradle factories becoming common in New York before the middle of the 19th century. The grain cradle was a big improvement as it permitted a man to stand erect while harvesting and the grain stalks were laid straight in small piles. A good man could cradle as much as 4 acres of standing grain in a day. After the grain had been cut, several strands of grain were wrapped around a small bundle and twisted at the end to form a knot holding the bundle together. Women often did the binding immediately behind the cradler. To dry the grain, bundles were set upright in shocks with their butts on the ground.

Threshing was accomplished using a flail. The grain bundles were laid out on the floor of the barn and repeatedly pounded with the end of the flail to remove the kernels of grain from the straw. The straw was then removed and the grain and chaff scooped up with a shovel, and placed on a winnowing board. The farmer winnowed the grain on a day when there was a good breeze. He opened the doors on both sides of his barn, tossed the contents of the winnowing board up in the air and the wind carried the chaff away as he caught the grain on the winnowing board. The grain was then placed in sacks woven of the tow, which were the shorter fibers from flax not as desirable for clothing, and stored until it was transported to market.

Hay was harvested around the 4th of July by cutting with a scythe. It was raked by hand with a bull rake about four feet wide. The flop-over rake, which preceded the dump rake and was pulled by a horse,

was invented in 1824 and eliminated a great deal of hard labor.

Farming was extremely labor-intensive in the period prior to 1825. There were three beasts of burden: ox, horse and man. Farming provided a livelihood for many thousands of New York families, giving most of them independence from the demands of an employer and an opportunity for success. Mechanization was beginning to come to agriculture but it would be a gradual process over several generations.

Apples, potatoes and squash were an important part of the farm family's diet and were often stored in a root cellar for consumption during the winter. Many of the early homes did not have cellars so a hole was dug in a well-drained spot on the side of a knoll and lined with straw, small brush or grass. The produce was placed in the depression and covered with similar material and then some sod. The settlers periodically made a small opening in the cover to retrieve whatever they needed.

The settlers had learned from the Native Americans how to tap the maple tree for its sweet syrup. New York is blessed with multitudes of maple trees and the settlers put this knowledge to good use. Cane sugar was expensive and often unavailable, making maple sugar a precious commodity. The settlers that had access to the sugar maple tree produced maple sugar for their own use and often for sale to others. In 1855, almost five million pounds of maple sugar were produced in New York. Maple syrup was not produced in any quantity until after the middle of the 19th century when containers became available for its storage.

Maple sugar from the New World had become famous in Europe in the late 1700s and there were people who dreamed of supplying the world with sugar from the maple tree. An example is the Holland Land Company that sent Gerrit Boon to New York in 1791 to acquire land and to set up sugar production. The company purchased 30,000 acres near Rome and later, two blocks of 45,000 acres and 65,000 acres near Black River, hoping to produce 1½ million pounds of maple sugar a year. Boon set up an experiment on 17 acres but produced less than $3,000 of sugar and the experiment died.[6]

6   Hedrick, Ulysses Prentiss A History of Agriculture in the State of New York p. 148-149

Filling the woodshed was a common, late summer or early fall task on almost every farm. In the 19th and early 20th centuries, sleds were commonly used instead of wagons for carrying heavy loads during the entire year.

The other source of a sweetener for the settlers was honey. The honeybee was not native to North America but was imported from England in the 17th century. The bees multiplied and became important for the production of honey and for pollination. Amazingly, the honeybee spread in the wild as well as domestically, especially since salvaging the honey destroyed the old-fashioned straw beehive along with its bees. It was not until the early 1850's that the movable frame hive, similar to what we use today, was introduced. It allows the harvest of honey without destroying the bees.[7]

Nature provided a number of minor sources for settler income in various parts of the State. Settlers on Long Island sold oysters and clams while settlers along the Hudson sold shad and those along Lake Erie, Lake Ontario and the Finger Lakes caught and sold trout and salmon. In the newly settled areas wild game was trapped for the sale of furs and ginseng roots were harvested in the forests. A ship that sailed from New York to China in 1785 carrying ginseng, received the unbelievable price of from $100 to $300 an ounce. When word of this reached

the United States, the woodlands from Maine to Florida were scoured for the roots.[8] Limited quantities of ginseng available in the forests permitted the price to remain high and a number of New York farmers produced ginseng in special shaded buildings during the 19th century.

Fish, wild game and native plants supplemented the food of the settler, but often there were serious periods of hunger. In 1789, there was a critical shortage of food in Otsego County with grain prices also high in Albany. Judge William Cooper of Cooperstown obtained a grant of 1,700 bushels of corn from the New York State legislature, packed it on horses' backs and transported it to needy families. In 1794, food was short in Tioga and Chemung Counties because of the large influx of new settlers. Examples of food shortages similar to these could be found in almost any of our counties during the early years of settlement.[9] The year 1816 became known as the year without a summer. Ash from a volcano in the Far East limited the quantity of sunshine reaching the earth and in Upstate

7   van Wagenen, Jared Jr. *The Golden Age of Homespun* p. 175-178

8   Hedrick, Ulysses Prentiss *A History of Agriculture in the State of New York* p. 136
9   Ibid. p. 101

New York there was a frost every month during the spring and summer in many locations. Crops were very light and thousands of settlers had a difficult winter scouring the countryside to find food for sustenance.

Disease was another enemy of the settlers, especially in areas where mosquitoes bred in stagnant ponds of water. The settlers called the disease, which today we know as malaria, such names as "Genesee Fever" or "the Ague". They didn't know the cause of this fever that resulted in much sickness and many deaths. Medical knowledge was limited and doctors were almost non-existent until a community was well-settled.

Since there were no roads or canals for the early settlers to travel, villages arose along the waterways of the State. The waterways provided paths of travel and served as a source of waterpower for the first sawmills, gristmills and other early industries. Before roads and canals were constructed, the best way to transport crops was by boat down the Susquehanna to Baltimore or down the Mohawk to Albany and the Hudson to New York City. Oswego became a marketing location after the British abandoned Fort Oswego in 1796.

The housewife's role was fully as demanding as her husband's with a multitude of tasks to perform. In addition to cooking, maintaining the home, caring for the children, growing some vegetables, milking the cow and making butter and cheese, she spun the wool and flax and then wove it into clothing for the family. **Table A,** in the Appendix, shows the yards of cloth made by families in the home in 1810. The population of the State was less than one million and well over eight million yards of cloth were made in New York homes that year. Young girls in the family learned to spin and weave at a very young age to help meet the needs of the family. The wool had to be washed and carded before spinning and the flax had to be grown and processed. Ready-made clothing, much of it from England, would have been available in New York City but was very expensive and not available upstate.

Flax represented over five million yards of the cloth made in 1810 and most of the early settlers had a plot of flax for the family's needs. The acreage grown is not available until 1845 when the New York census listed 46,000 acres, but the acreage was likely

to have been greater earlier in the 19th century. It was pulled rather than cut because the fibers extended into the roots. After harvest it was retted (rotted) to break down cellulose around the fibers. Next it was dried and went through a flax brake followed by swingling and hatchelling, all to remove the cellulose around the flax fiber. The shorter fibers, called tow, were not as desirable and were used to make sacks for grain. The longer fibers were turned into clothing that was durable but did not provide much warmth in cold weather.

Although today, with our rapid means of travel, Suffolk County seems close to New York City, a look at information provided in the *1820 New York Gazetteer* shows us that agriculture and life on the farm, in Suffolk County, were quite similar to the other parts of the State. There were 4,642 persons employed in agriculture and there were 175,994 acres of improved farmland in the county. There were 22,902 cattle, 36,607 sheep and 4,498 horses. In addition, there were 64 gristmills and 42 sawmills with creeks powering some of the mills. Some salt was also produced from the salt water readily available. These numbers are consistent with the numbers from many of the other counties. Land was still being cleared and the agriculture was a combination of livestock and grain farming.

The 1820 *New York Gazetteer* also shows that Queens County, which later was divided to form Nassau County on the western end of Long Island, also was an agricultural county with 4,130 persons out of a population of 21,519 whose trade was agriculture. It listed 118,022 acres of improved farmland, 14,457 cattle, 21,357 sheep, 5,282 horses, 44 gristmills and 21 saw mills. The tide in the bays, coves and inlets supplied power to many of the mills.

The first settlers in an area, of necessity, relied totally upon their own resources. But as more settlers came to any area, developing services needed by others evidenced the entrepreneurial spirit of many. Gristmills and sawmills usually came first followed by the blacksmith and the general store. An ashery solved the transportation problem of bulky ashes and a tanner was essential to provide shoes, harness and other leather goods. In the early years of a settlement tanning was a seasonal part-time occupation, since most of the butchering was done in the fall and winter when the cold weather preserved the meat. An interesting by-product of tanning was the production of cow hair that was mixed in plaster for walls of houses. The cow hair helped hold the plaster together in much the same way as fiber put in concrete does today. In the tanning process the hides were put in a vat of lime for soaking to loosen the hair. **Table B**, in the Appendix, illustrates the significance of the 1810 tanning industry in New York. There were 867 tanneries processing 151,165 hides, 210,445 calfskins and 61,616 sheepskins. The Census Marshall placed a value of a little over a million dollars on these hides.

A family on a well-traveled road often opened a tavern and soon there would be a preacher, a doctor, a schoolteacher, a builder, a cabinetmaker and an undertaker. Small manufacturing establishments appeared producing plows, grain cradles, nails and a variety of the other needs that had been present in the developed communities the settlers had left. These services moved into new communities very rapidly after settlement, but most early settlers survived on their own talents until the population was sufficient to support these services.

Chapter V
Roads, Canals and Railroads

ROADS

Little regard was given to the need for roads in Colonial New York with reliance instead upon natural waterways as avenues of travel. Upon becoming a state, the New York Legislature provided grants and held lotteries to raise money for construction of roads but found these methods inadequate to meet the needs. The State then devised a plan to use private money to build roads for travel and the transportation of produce from farm to market.

The first road constructed under this proprietary program was the New York and Albany Post Road on the east side of the Hudson River, extending from New York City to Albany in 1785. Exclusive rights to run stages on this road were provided by the State, with stiff fines for infringement. Stages ran twice a week making the trip in three days the first year, but in the next year it made three trips a week taking only two days a trip. It remained profitable until 1812 when it was turned over to the State because of competition from steamboats on the Hudson River.[1]

Albany, located on the Hudson and close to the Mohawk, became the center of transportation with turnpikes going in all four directions. The first road west of Albany, called the Mohawk Turnpike, was

---

1 Hedrick, Ulysses Prentiss *A History of Agriculture in the State of New York* p. 171-173

A turnpike tollgate near Manlius in 1899. There were hundreds of tollgates in the State, in the 1800s, where private builders of roads and bridges collected fees. Courtesy of the OHA Museum & Research Center.

Albany & Bethlehem Turnpike Road/Rate of Tolls. Courtesy of Albany Institute of History & Art, u1972.16.1.

built in 1793, and extended from Albany to Schenectady. It was constructed of broken stone, was raised sixteen inches in the center, with sixty feet of width and sloped to the edges. Three years prior to the construction of this highway, which is now Route 5, a stagecoach had been traveling the same route with passengers paying three cents a mile.[2]

The Mohawk Turnpike was extended up the Mohawk to Canajoharie with stage service continuing to Geneva in 1797. In 1807, a charter was granted for a stagecoach to Buffalo, but it wasn't until 1811 that stagecoach operations connected Albany and Buffalo. The Post Road from Albany to Buffalo through Utica, Auburn and Geneva was a main artery for travelers heading to the west even after the construction of the Erie Canal. The charter name was "Seneca Turnpike" but it was commonly referred to as "The Genesee Pike" and often called "The Ohio Trail" because of the multitudes traveling on it to Ohio. With almost 300,000 settlers in Western New York by 1810, we can imagine the conditions of travel they encountered to reach their various destinations.

A bridge over a mile long, called an engineering marvel, was constructed across the northern end of Cayuga Lake and completed in 1800. It burned in 1804, was replaced in 1812 and abandoned in 1857. The bridge was constructed in 18 months and cost $150,000. It saved many miles of travel by eliminating the need to go around the Montezuma marshes.[3]

As the number of settlers moving up the Hudson Valley and on up the Mohawk Valley increased, the need for improved roads became evident. Gradually roads were constructed to fill the existing need. The 448 miles of highway required to extend from New York City to Buffalo was achieved with four major roads; the Albany Post Road of 149 miles, the Mohawk and Hudson Turnpike of 15 miles, the Old Mohawk Turnpike (Schenectady to Utica) of 80 miles and the Seneca Turnpike of over 200 miles.[4]

Ulysses Hedrick provides information from the 1807 *"Transactions of the Society for the Promotion of Useful Arts"* by Benjamin DeWitt listing 88 incorporated turnpike road and bridge companies with contracts for over 3,000 miles of turnpikes and 20 large toll bridges in New York State. The Capital stock of these companies varied from about $1,000 to $3,000 for each mile of road they were to construct. The total Capital stock of these companies was over five million dollars. There were tollgates at bridges and about every six miles along the turnpikes. Although five million dollars was a very large sum of money in 1807, when we consider that it represented over 3,000 miles of road it is obvious that the roads were primitive compared to our roads today.

By 1845, New York, with 457 turnpikes, had more turnpikes than any other state. The greatest numbers were formed in the first and second decades of the 19th century with 126 and 133 respectively. Plank roads became the rage beginning in 1846 with the construction of a plank road north of Syracuse, from Salina to Central Square. There

2   Charles B. Knox Gelatin Co. Inc, *Old Mohawk Turnpike Book*

3   Hedrick, Ulysses Prentiss *A History of Agriculture in the State of New York* p. 178

4   Charles B. Knox Gelatin Company *Old Mohawk Turnpike Book*

The first of hundreds of plank roads in the United States, circa 1900. The loaded vehicle had the right of way, requiring the empty vehicle to go off the road on the dirt. Note the mile marker by the side of the road. Courtesy of Mr. & Mrs. Harold Baker.

were a total of 335 plank roads constructed in New York but they gradually fell out of favor because of their high maintenance cost.[5]

Early road building consisted of cutting the necessary trees and removing the stumps. An ordinary plow was used to plow dirt from the outside of the road toward the center. Stones were thrown into the depressions along the side of the road. This procedure was followed year after year and provided poor support for the wheels of stagecoaches and wagons traveling the road. The establishment of turnpikes improved these roads but on less-traveled roads country road maintenance followed this procedure into the early 20th century. Corduroy roads were constructed in wet areas by cutting trees and laying the trunks on top of each other until they reached above the surface of the ground. This type of road was extremely rough and difficult, especially for passengers in stagecoaches. Dewitt Clinton describes freight traffic in a journal of a trip he took to Western New York in 1810:

"The large wagons carrying 40 to 50 hundredweight, go from Geneva to Albany for about $3 a hundred,

A typical country road in many parts of the state, circa 1920.

5 Klein, Daniel B. & Fielding, Gordon J. *Learning from the 19th Century; Private Toll Roads*

carrying and returning with a load, which makes about $6 a day, as they consume 20 days out and home. They make 13 trips a year and find it profitable two-thirds of the time. They generally use five horses; the rims of the wagons are six inches broad, and one has nine inches and six horses."

Roads came into Northern New York more slowly than many parts of the State. The St. Lawrence, with Montreal providing an excellent market for crops, minimized the need for turnpikes. Northern New York grew and prospered until after the Erie Canal was constructed, which brought large harvests to market from the inexpensive and rich soils in Ohio and other mid-western states. This competition made it difficult for Northern New York farmers, with its less productive soils, to compete in the marketplace.

It was not uncommon for the path of a road to be determined by a few bottles of whiskey or the offer of help in the road's construction. Almost every farmer wanted the road to pass in front of his farm to make both travel and marketing of crops more convenient. Many of our New York roads are winding and full of curves because it was easier for horses and oxen to pull their loads around hills and wet areas.

CANALS

When we think of canals in New York we think of the Erie Canal, but work done by the Western Inland Lock Navigation Company, formed in 1792 with General Philip Schuyler as President, was of significant importance in developing New York's canal system. Short canals were constructed near Fort Herkimer, Little Falls and Wood Creek creating navigation without portage from Schenectady to Cayuga Lake, the Oswego River and Lake Ontario. The costs of construction were far greater than anticipated so the company was not a financial

An engraving of "The Clermont", the first commercially successful steamboat. This ushered in a new age for the transportation of agricultural products. Photo courtesy of New York State Archives.

The arrival of the "Seneca Chief" in Albany from Buffalo, on November 2, 1825 with Governor DeWitt Clinton and other dignitaries on board; part of the ceremonies marking the completion of the Erie Canal. The Erie Canal, coupled with the New York Harbor, made New York the "Empire State". Courtesy of the New York Museum of Cheese.

success, but a great deal was learned that was later beneficial in the construction of the Erie Canal. Via these short canals, several portages were eliminated and freight could be shipped as far west as Seneca Falls and farm products shipped from that area to Schenectady where they were transported by land to the Hudson River at Albany. The water level of the Mohawk River at Schenectady, 15 miles from Albany, was 211 feet higher than the level of the Hudson River at Albany, making transportation on the river treacherous and foolhardy until the 1825 construction of the Erie Canal, with its five locks between the two cities.

Prior to the construction of these short canals, flat-bottom boats, called bateaux, with little draft and powered by oars and poles, carried freight and passengers from portage to portage along the Mohawk River to Oneida Lake. It was a slow process and hard work to transport product over the portages. The small canals, some of which were several miles long, substantially reduced the cost of transporting a ton of farm produce from Seneca Falls to Schenectady. The Dutch founded Schenectady in 1661 at the lower navigational limit of the Mohawk, where all travelers and freight going in both directions passed.

Robert Fulton's steamboat, the Clermont, issued in a new age in transportation in 1807 when it traveled up the Hudson River from New York City to Albany at an average speed of 5 miles an hour. Travel up the Hudson to Albany by boat had been subject to the whims of nature, with one never knowing whether the trip would take a couple of days or a week. The steamboat's success had little immediate effect upon shipping of agricultural products, but became more important in the succeeding decades, not only on the Hudson River but on the many other inland waterways of New York. The first steamboat on our inland lakes was in 1819 on Cayuga Lake. For the next century there were numerous steamboats on our large inland lakes including the Finger Lakes, Oneida and Chautauqua lakes as well as Lake Erie, Lake Ontario and the St. Lawrence River. These waterways served as the highways for the marketing of crops from thousands of New York farmers located near their shores. For these farmers water transportation was more economical and faster than using the poor roads existing in the countryside. The steamship also opened additional opportunities for marketing more perishable crops with faster, cheaper and more dependable service from upstate New York to New York City and on to Europe.[6]

These steamships were powered by wood from the forests of New York for many years. Although coal became available in the 1820s, wood was the choice for steamships until many years later and provided an additional source of income for farmers near our waterways. In 1825, the steamboats on the Hudson and the ferryboats at New York City consumed 3,000 full cords of wood each week. Add to this the great quantities consumed on our other lakes and one can appreciate the significance of wood as a source of income for New York farmers.[7]

The following is a partial account of lake traffic from the Buffalo *Commercial Advertiser* in 1847.

> *"There were in commission on the lakes, 98 steamers, 495 schooners and numerous other boats with a total tonnage of 131,460. Some of the prominent items arriving in Buffalo that year were; 1,857,000 barrels of flour, 6,489,100 bushels of wheat, 2,862,000 bushels of corn, 8,800,000 barrel staves, 3,436,000 pounds of lard and 42,000 barrels of pork."*

> *"There were exported from Black Rock and Buffalo, by canal in 1847; 710,943 tons, principally the products of field and forest, of the regions bordering the western lakes. The total value of imports of Buffalo from the lakes, in 1846 was ascertained and estimated to amount to nearly $20,000,000. In the same year there arrived at Buffalo, via the Erie Canal, the great bulk of which was shipped to the west, 153,761 tons of merchandise and other property valued at $23,199,665. The number of passengers arriving and departing from Buffalo in 1846, was not far from 250,000."*[8]

The numbers listed above give us a good indication of the traffic moving through Buffalo on the Erie Canal. A large quantity of the goods traveling west through Buffalo was undoubtedly produced in New York State.

The possibility for a canal from Albany to Lake Erie was suggested by a variety of people beginning in the latter part of the 18th century. An assembly committee was formed in 1808 to consider a canal

and a survey was ordered to determine its viability. James Geddes, of Onondaga, made the survey and reported that a canal from the Hudson to Lake Erie could be made without serious difficulty. In 1811, a bill was passed to carry out the project and the national government was invited to participate. Washington had interest in other projects and it became evident that no help would come from that direction. A commission had been appointed and was authorized to borrow money and take cessions of land for the canal but not to proceed with its construction. The Holland Land Company offered over 100,000 acres for the canal's passage through its property. Others also offered grants of land while some attempted to obtain heavy damages for its passage through their property. The War of 1812 held up action to move ahead with the canal but construction was authorized by the legislature in 1817 through the efforts of DeWitt Clinton. The first ground was broken at Rome on July 4, 1817. The authorization for construction of the Erie Canal estimated the cost to be $4,571,813 and the authorization, at the same time, of the construction of the Champlain Canal was estimated to cost $871,000.[9]

Much of the original Erie Canal was dug by farmers under contract with New York State, especially sections of the Canal adjacent to their farmland. The farmers were pleased to have the Canal adjacent to their property and digging a portion of the Canal provided the farmer with a source of additional income.

The Canal opened in sections. In 1819, the section between Utica and Rome opened and the 63-mile-long Champlain Canal was completed. The 363-mile-long Erie Canal was finally completed from the Hudson River to Lake Erie with a big celebration on October 26, 1825. Some construction continued until 1836 making the total cost a little over seven million dollars. The canal accommodated boats up to 78 feet long, 14½ feet wide and with a 3½ foot draft capable of carrying a 75 tons load.[10]

There were many opposed to the Erie Canal because of the cost and the general feeling that it could not be successfully completed. DeWitt Clinton, who ran for Governor of New York State on a pro-canal

6   Hedrick, Ulysses Prentiss *A History of Agriculture in the State of New York* p. 240
7   Ibid. p. 241-242
8   Ibid. p. 239-240
9   Roberts, Ellis H. *American Commonwealths, New York* p. 533-535
10   Ibid. p. 537-542

Changing tow horses pulling an Erie Canal boat, courtesy of the New York Museum of Cheese.

Left: DeWitt Clinton, promoter of the Erie Canal.

Right: An 1834 sign promoting travel on an Erie Canal packet boat in conjunction with the more rapid train service from Albany to Schenectady. Courtesy of the New York Museum of Cheese.

DEWITT CLINTON
became the Erie Canal's strongest supporter

1834.
PACKET BOAT,
AND
Rail-Road
ARRANGEMENT.

A Packet BOAT will leave Schenectady
Daily, for Utica, Rochester and Buffalo, at

Half past 10 o'clock A. M. and
Half past 6 o'clock P. M.

PASSENGERS for the PACKETS will leave Albany
by the CARS at 9 A. M. and 5 P. M.

These are the only Cars that run to the Packets.

By this arrangement there is no delay, as the Packets will
leave Schenectady immediately after the arrival of the Cars.

platform, was always an avid supporter of the canal. Detractors referred to the Erie as "Clinton's Ditch" and "Clinton's Folly". It turned out to be a great success from the very beginning with tolls, even before it was finished, exceeding the interest cost on the borrowed money.

It is difficult to comprehend the tremendous effects of the Erie Canal on New York and its agriculture. The canal made large cities out of small villages and thriving villages appeared out of nowhere. New York became the Empire State and New York City the metropolitan city of the western hemisphere. It developed Buffalo into a large shipping center, Rochester became known as the Flour City, Syracuse the Salt City and it spawned dozens of thriving cities and villages along its route. Before the construction of the canal the average cost of shipping a ton of freight from Buffalo to Albany was $88. Even before the canal was totally completed the cost of shipping dropped to less than $22 and by 1835 the cost of shipping was down to $6.44 a ton.[11]

In 1826, the first full year of their operation, 19,000 boats and rafts passed through Troy on the Erie and Champlain canals. Canal boats, of necessity, had to be towed up and down the Hudson River if their cargo was going on to New York. Steamships charged from $5 to $15 to tow a boat depending upon the competition. Normally there were 60 to 80 canal boats in a tow with the record being 108.

Wheat, in Western New York, became worth four times what it had been previously. Before construction of the Erie Canal, the best way for Western New York farmers to get their produce to market was to Bath on the Susquehanna. In 1800, a bushel of wheat in Geneva worth 60 cents was worth $1 in Bath. Upon the completion of the Erie Canal the products from western New York farms moved east on the Erie rather than south to Bath. Farmers in New England had shipped potatoes to Albany but with the Erie Canal's completion they could no longer compete with Central New York potatoes. New York had ranked second to Pennsylvania, among all the states, in the production of agricultural products, but from the time of the opening of the Erie Canal until after the Civil War, New York ranked first.[12]

Figures taken from *The Northern Traveler* published in 1834, illustrate the importance of the canals to our state:

"The whole quantity of down freight upon which toll is charged by weight, that was conveyed on the New York canals to Albany in 1833, amounted to 152,945 tons. Arrived, 734,133 barrels of flour, 22,922 barrels of ashes, 13,489 barrels of provisions, 19,908 barrels of whiskey, 873 hhds. of whiskey, 17,116 barrels of salt, 298,504 bushels of wheat, 122,944 bushels of course grain, 257,252 bushels of barley, 2,187 boxes of glass. And also the following upon which the toll is not charged by the ton: 20,960 cords of wood, 74,350 feet timber, 55,338,547 feet lumber, 74,350 M shingles, and 68,321 tons of merchandise, furniture and sundries sent up the canal from Albany. The whole amount of toll received by the collector at Albany is $323,689.88 making an increase of $87,053.56 over the receipts of last year. The number of boats arrived and cleared was 16,834."

In 1837, a half a million bushels of wheat were shipped down the canal from Buffalo and four years later it reached a million. Prior to construction of the canal, New York City was the nation's fifth largest seaport, behind Boston, Baltimore, Philadelphia and New Orleans. By 1840, New York was the busiest seaport in America, moving tonnages greater than Boston, Baltimore and New Orleans combined. After being open only nine years the canal recouped in tolls the entire cost of initial construction.[13]

The success of the Erie and Champlain canals was the beginning of a building craze for additional canals in the State. The Chenango connected the Erie Canal at Utica with the Susquehanna River at Binghamton. It was 97 miles long with 14 miles of feeders. The Black River Canal connected Rome with Boonville and then on to Carthage. The Oneida Lake Canal connected the Erie with Oneida Lake and the Oswego Canal connected Syracuse with Oswego. Cayuga, Seneca and Keuka lakes were connected to the Erie and the Genesee Valley Canal ran from Rochester to the Allegany River at Olean. There were numerous other canals that also provided arteries to markets for the farmers of New York.[14]

---

11   Roberts, Ellis H. *American Commonwealths, New York* p. 544
12   Hedrick, Ulysses Prentiss *A History of Agriculture in the State of New York* p. 247-248
13   New York State Canal System
14   Hedrick, Ulysses Prentiss *A History of Agriculture in the State of New York* p. 249-252

Erie Canal Aqueduct over the Genesee River at Rochester.

Deepening and widening the Erie Canal, courtesy of the New York Museum of Cheese.

The average toll on the canal in 1839 was $1.12 a ton, which was gradually reduced over time to 12 cents. Tolls were totally abolished in 1882 when the railroads were taking away the business on the canals. In 1862, the tolls collected on the canals totalled over five million dollars and the maximum tonnage carried on the canals was over six million in 1868.[15]

The immediate success of the Erie Canal soon brought need for improvements and enlargement. An enlargement was ordered in 1835 and completed in 1842 at a cost of $23,000,000. In 1847, enlargements began again which were completed in 1862. Another enlargement was ordered in 1884 including the lengthening of the locks. Finally, in 1903, the State changed the canal's name to "The Barge Canal", authorized an expenditure of $140,000,000 for another enlargement and made significant changes in its route. As the tonnage on the canal increased there was need for larger boats and the canal was in continual need of enlargement. The canal gradually evolved from 42 feet in width and 4 feet deep to 200 feet wide by 12 feet deep when the Barge Canal was completed.[16] The faster railroads and the paving of highways in the 20th century gradually reduced the freight tonnage on the canals and today the remaining ones serve primarily as recreational waterways.

The opening of the St. Lawrence Seaway in 1959 was heralded as a major milestone in transportation to and from the states adjoining the Great Lakes. The Seaway has had little effect upon agriculture in New York even though we have ports at both Oswego and Buffalo. Some grain moves in and out through these ports but has little significant effect.

Farmlands close to the canals became more valuable and these farmers tended to prosper while those removed from ready access to the canals found it difficult to compete and their property decreased in value. Once thriving villages like Bath and Catskill found themselves removed from the main artery of transportation and commerce and withered in a localized depression. The completion of the canal also marked the abandonment of many small and marginal farms in the New England states. Produce from the rich soils of western New York,

Ohio, Michigan, Indiana and Illinois could now be shipped to Albany and on to New York at a price less than the cost of producing them in New England. Farmers in New York that had been supplying the Albany and New York markets with grain soon found it difficult to compete with the lower prices created by the influx of western grain and looked to alternative crops to bolster their income while those on small less productive farms found other means of employment. Farmers gradually began to move toward the production of crops that were more difficult to transport at greater distances from markets. The Erie Canal marked the beginning of a process that continues today, where price pressures from outside the State often determine changes in New York's agriculture.

RAILROADS

In 1826, the year after the Erie Canal was completed, the first railroad in the country to be given a charter was the Mohawk and Hudson in New York. Work on the railroad did not begin until 1830 and it was on August 9, 1831 that a tiny, four-ton engine, named the DeWitt Clinton, drew three coaches from Albany to Schenectady. The rails were ½ inch thick strips of steel, two inches wide, spiked to four-inch square timbers, laid on well-spaced ties. Smoke and sparks flew from its stack scaring both children and horses as well as setting passengers' clothing on fire. The coaches were adapted from stagecoaches traveling on the turnpikes.[17]

The Erie Canal was already in existence between Albany and Schenectady so the railroad wasn't needed to carry freight. The canal traveled a circuitous route with several time consuming locks making the railroad a timesaver for people traveling between these two cities. For many years the State wouldn't permit a railroad to be located parallel to the Canal as they didn't want to lose any toll revenue and therefore designated railroad locations to serve as feeders to the canals.

A similar situation existed in western New York where the distance between Rochester and Buffalo was 85 miles via the Erie Canal and where a number of locks made the trip time consuming. The Rochester and Tonawanda railroad was formed in 1836 shortening the distance between the two cities to

15 Roberts, Ellis H. *American Commonwealths, New York* p. 543
16 Hedrick, Ulysses Prentiss *A History of Agriculture in the State of New York* p. 248
17 Ibid. p. 256

The First Steam Train in America. Mohawk & Hudson Railroad excursion train from Albany to Schenectady in 1831. The first cars were converted stagecoaches designed to run on rails. Courtesy of Albany Institute of History & Art.

70 miles and substantially decreasing the length of passenger travel time.

The Mohawk and Hudson Railroad was the beginning of what later became the New York Central railroad system. It was extended to Utica in 1836 and to Syracuse at about the same time. About a decade later, the railroads in the eastern part of the state were connected with those in western New York, opening another great artery for transportation and commerce from the mid-west to Albany. It wasn't until 1851 that a railroad reached Albany from New York City, partially because of the difficult terrain and partially because powerful steamship interests were opposed to it.[18]

Numerous railroads were chartered throughout the state during the 1830s and 40s with each one operating separately from the others. In addition, track width had not been standardized so each railroad company decided their track width. This required the physical removal of freight from one line to another at the terminus of each railroad line. Because of this situation shipping goods by railroad was expensive and slow whenever shipments involved two or more railroads. Shipping by rail, under these conditions, limited the shipment of perishable farm products.

In 1853, the New York Central Railroad Company was organized, consolidating nine smaller railroads, resulting in dramatically increased transportation efficiency. The State, as a condition to permitting the consolidation of these railroads, required the railroads to pay the State the same fees on the freight they carried as would have been paid if it had been traveling on the canal. That requirement was soon repealed by the State. The State also prohibited farmers from shipping their products on the railroads during the seasons of the year that the canal was operating, even for perishable crops, which might be spoiled by the time markets were reached on the much slower canal. This was a contentious rule for farmers and eventually they succeeded in gaining its repeal.

A railroad was proposed to pass through the southern tier counties in the 1830s. Most of this area of the State had missed the advantages of canal transportation that the northern counties had received from the construction of the Erie Canal. Due to insufficient interest by investors, it wasn't until 1851 that the Erie Railroad was completed from New York City to Dunkirk on Lake Erie. Upon its completion, the Erie Railroad became the most direct route from Lake Erie to New York City. The Erie Railroad provided markets for farmers with a much lower cost of transportation than was previously available. It

---

18 Hedrick, Ulysses Prentiss *A History of Agriculture in the State of New York* p. 257

A grain elevator of The New York Central and Hudson River Railroad from an 1877 sketch in Harper's Weekly. This was prior to the merger of the two railroads. Note both canal boats and sailing ships in the harbor and the train in the elevator.

also provided market access for the huge quantities of timber in the Southern Tier counties. The Erie Railroad was largely responsible for the increased growth of Binghamton, Elmira, Corning, Hornell, Olean, Salamanca and Jamestown and made good markets for farmers surrounding them. It was a two edged sword, however, for New York's agriculture as a whole, because it now brought mountains of grain and great quantities of livestock from Ohio and Michigan within three days of New York City, the market metropolis of the United States.[19]

Corn in the West sold at from 10 to 30 cents a bushel whereas in the East it brought 75 cents to $1.00 a bushel. The railroad carried livestock daily to Eastern markets at prices below the cost of eastern production. Farmers in New York could only afford to market by-products of the dairy industry to the butcher: dairy cows and working oxen.[20]

Prior to the 1850s, railroads successfully carried thousands of passengers due to their faster speed, whereas the Canals continued to carry the bulk of freight due to their lower shipping cost. Once the railroads consolidated in the 1850s and 60s and were able to lower freight costs, they gradually carried a larger portion of the freight than the canals. In 1843, seven railroads connecting the larger cities of the State received $590,000 in passenger fares and $190,000 from freight.[21] Data from 1885 shows over 25 million tons of freight carried on the railroads whereas slightly less than 5 million tons was carried on the canals.[22]

In 1840, New York had 363 miles of railroads, increasing to 1,361 by 1850 and to 2,682 miles in 1860. Railroads were bringing markets to thousands of our New York farms as railroad depots became located within a reasonable travel distance

19 Hedrick, Ulysses Prentiss *A History of Agriculture in the State of New York* p. 255-258
20 Ibid. p. 266-267

21 Ibid. p. 260
22 Roberts, Ellis H. *American Commonwealths New York* p. 642

from the farms.[23] Cheese factories began to appear, permitting a uniform product to be manufactured from the milk of many farmers who had previously produced cheese on their farms; cabbage and potatoes could now be shipped to more distant markets; and flourmills could purchase wheat from local farmers to be milled and shipped to distant points.

Hedrick sums up the effect of railroads upon agriculture in the following excerpt from his book:

"Railroads were not all to the good for agriculture in New York. In the fifties and sixties, whether or not the railway in the eastern states was an asset or a liability to a farmer was a question for discussion. Rails to the Mississippi and beyond at this time opened up vast quantities of fertile and easily tilled land in the Middle West on which were transported wheat, corn and livestock to eastern markets at a price which was ruinous to the eastern farmer. Prairie agriculture was much cheaper than woodland agriculture. Farmers in New York and New England emigrated to Illinois and Illinois to homestead prairie land. Abandoned farms in the hills of New England and New York were plentiful as early as the 1860s. At the close of the Revolution, the Hudson River Valley was called "The Bread Basket of the Nation;" the next generation called the Genesee Country the "Granary of the Nation". The Erie Canal robbed the farmers along the Hudson of their supremacy as food providers; the railroads in their turn permitted the establishment of the Nation's granaries farther and further to the west. Moreover, railroads were drawing heavily on labor from farming communities for construction and operation. It was estimated in 1857 that the roads of the country were employing in one capacity or another 400,000 men, a large proportion of whom came from farms."

Railroads continued to serve as the major means of moving New York's agricultural products to market through the remainder of the 19th century and to the middle of the 20th century. Even with the enlargement of the Erie Canal, renamed the Barge Canal, the tonnage of agricultural products moving on canals continued to decrease while railroads took a larger share. After World War I the internal combustion began to again change the methods of moving farm products to market. Concrete highways were constructed from city to city and to villages that were usually within a few miles of most of our farms.

These highways were maintained for year round traffic and trucks began to move farmers' products to market replacing the horse and wagon. The volume of freight carried on the railroads was not appreciably affected until after World War II when super highways were constructed for use by both passenger cars and large tractor-trailers. With these super highways, trucks could now compete quite effectively with the railroads in moving farm products more quickly to markets within a few hundred miles. In addition, the tractor-trailer can go directly to the farm, load the product on the truck and deliver it directly to the market with no handling inbetween.

Railroads continue to serve the farmers of New York today but in a more limited capacity. They bring grain to New York from the mid-west but now it is often to feed dairy cows or bring corn to make ethanol, and not for shipping from the port of New York City. The railroads also are intermediaries between trucks on each end of the shipment. Trucks deliver loaded trailers to the railroad for long distance shipments and trucks pick the loaded trailers up from the railroads at the other end. The direct value of this method of shipping to New York's agriculture is quite minimal.

The transportation of agricultural produce in New York has undergone great change since 1784; sometimes beneficial and sometimes detrimental to the interests of New York's agriculture but generally a mix of the two. Transportation of agricultural produce by airplane continues to follow this "mixed blessing". Fresh cut flowers arrive overnight from countries in South America at a price that is lower than the production costs of New York producers. With the innovations in transportation, all parts of the world can market their agricultural products in New York and conversely New York can market its agricultural products anyplace in the world.

---

23   Thies, Clifford F. *The Cato Journal Fall 2002* p. 19

Chapter VI
Mechanization of New York's Agriculture

Mechanization of agriculture moved forward slowly for thousands of years from the time man first planted seeds and harvested crops from them some 10,000 years ago. Crude stone and wooden tools were used exclusively until the Bronze Age of about 5,000 years ago, when man devised a sickle made of bronze to aid in the harvest of grain. With the domestication of animals an ox was used to pull a forked stick for breaking up the soil and was hooked to a sled for pulling a load. The Iron Age of approximately 3,000 years ago brought improvements with more durable iron hoes, spades and sickles. It wasn't until about 1760, on the eve of the American Revolution and the Industrial Age, that iron began to be used more commonly in making new and improved agricultural tools. Initially these improved tools included the hoe, shovel, spade, axe and sickle, which were in common use at the time New York became part of the United States.

Wagons were uncommon on farms until well into the first quarter of the 19th century. Log boats pulled by an ox and fashioned from the "Y" of a tree, with a few split planks across the crotch to carry the load, sufficed for taking grain to a mill, hauling hay and taking the family to church on Sunday.

Hay was cut with a sickle, pulled together with a handmade wooden rake and pitched onto the log boat with a wooden fork cut by the farmer from a forked branch of a tree. Grain was cut with a sickle, bound by hand, threshed with a flail and cleaned by tossing grain and chaff from a winnowing board into the air and letting the wind blow away the chaff while the kernels of grain were caught on the winnowing board.

After hoeing, spading or plowing, an ox dragged a large tree branch, with multiple limbs, to level the ground. Grass seed and small grains were broadcast over the ground by hand and an ox pulling the tree branches would be used to cover the seed.

Ground was prepared in the same manner for planting corn, potatoes, beans and other row crops, but after leveling, a hoe or dibble was used to make an opening for the seed to be put in the ground by hand. An axe blade was sometimes used to make an opening in the ground for the placement of seed. Weed control consisted of hoeing and pulling weeds that were missed. Potatoes were dug with a spade and the ears of corn were husked by hand using a husking pin. The dry kernels of corn were removed from the ears by hand.

The large quantities of land that became available after the American Revolution in New York and states west of the Appalachian Mountains, along with the thousands of families settling on these lands, sparked an expansion in agriculture that was greater during the next century, than all that had happened in agriculture in the previous millennium. The Industrial Revolution started in Europe around 1760 and slowly moved to the United States, growing gradually until after the War of 1812 when it blossomed. Agriculture became a major industry at this time. With its significant needs for mechanization, inventive minds began to fill those needs. The 1850 census shows there were 135 manufacturers of agricultural implements in the State employing 923. By 1860, the number of agricultural implement manufacturers had increased to 333 with 2,904 employees.

Blacksmith shop, circa 1920. Every community had at least one blacksmith shop in the 19th century. The blacksmith not only shod horses but repaired and made all sorts of metal necessities for the farm. Courtesy of Kaye Forsythe.

Water attracted the early settlers not only for transportation but also for the power it provided for sawmills, gristmills and other early industries. Montville, in Cayuga County, was somewhat typical of many small developing villages. It had a barrel factory, a triphammer, harness factory, plow factory, distillery, scythe factory, gristmill, linseed oil mill, auger factory and a woolen mill. Between the time the Patent Office opened in Washington, D.C. in 1790, and November 30, 1880, Cayuga County inventors took out 68 patents on harvesters, 11 for carriage axles and boxes, 12 for plows, 10 for harvester knife grinders, 8 for threshing machines, 8 for churns and a number of others for such things as harrows, fence wire, fanning mills, animal pokes, and pumps. New machines and new methods were being devised throughout New York as well as the rest of the country.[1]

One of the novel inventions was an 1835 patent Henry Burden of Troy received for a machine that produced 60 horseshoes a minute. Previously, horseshoes had been heated, pounded out and bent individually by the blacksmith. It was a laborious task and generally produced a shoe that was inferior to the ones from Burden's new machine. Another significant innovation was the nation's first grain elevator, built in Buffalo in 1842 by Joseph Dart. It had a flexible bucket elevator arrangement that reached into the ship's hold removing the grain from the ship to the elevator.

One of the first seemingly small but significant labor saving inventions was the scythe, from which the grain cradle was soon developed. As early as 1812, the Waters Scythe Factory of Amsterdam, in the Mohawk Valley, turned out 6,000 scythes a year. The generally accepted date for the invention of the grain cradle was about 1776. The scythe had been put into use a few years earlier and some inventive mind decided to add several long wooden fingers to collect the grain stalks as they were being cut. The first US patent for a grain cradle was issued to Isaac Babcock, of New York State, in 1823. Soon grain cradles were being manufactured in dozens of communities in the State. An interesting example of the time needed to transfer technology a hundred years ago, is that the last patent issued for a grain

---

1   Hedrick, Ulysses Prentiss *A History of Agriculture in the State of New York* p. 290

clockwise from top: Farmer sharpening his scythe with a whetstone in a field of hay, circa 1900. Courtesy of OHA Museum & Research Center.

Loading piles of hay on a wagon in Washington County in 1913. Courtesy of New York State Archives.

Hay wagon with hay loader ready to load windrows of hay, circa 1920. Courtesy of Evelyn Hudson.

Baling straw in 2009. Each bale weighs almost one ton, so large fork lifts are used to handle the bales.

cradle in the United States was in 1924, over a half a century after the reaper had been in use!

The 1819 patent received by Jethro Wood for a plow with three separate cast iron parts: a moldboard, a landside and a share, was very successful with sales of almost 4,000 plows a year. It was a significant improvement over the plows that had been in use. Census figures are not available to know what proportion of our New York farmland was plowed prior to 1819, but as late as 1865 only about one-fifth of the improved farmland in New York was plowed each year. It is likely that an even smaller percentage was plowed prior to then, considering the plows available at that time. From New York's agricultural census of 1865 we do know that 242,236 farms were enumerated, reporting 3,058,012 acres plowed, making an average of 13 acres plowed on each farm.

In 1837, John Deere invented the self-polishing steel plow that was of greater importance on the prairie soils of the west than in New York. There were a number of plow manufacturers in New York including the Wiard family of Western New York, who, in 1876, established the Syracuse Chilled Plow Factory that developed a process of hardening cast iron to make a plow effective in our New York soils.

Inventive minds throughout the state devised harrows, first with wooden peg teeth and later with iron peg teeth, both of which were a great improvement over the previously used tree branch. About 1861, the spring tooth harrow appeared along with a variety of crude cultivators. The first cultivators were adaptations of the plow throwing dirt onto the row to cover weeds as well as to remove weeds between the rows. There had been attempts to devise successful seed drills by many people, including George Washington at Mt. Vernon, but credit for one of the first successful drills goes to Henry W. Smith in the 1850s.

Corn was planted by hand in New York well into the 20th century. At first farmers dropped the seeds from their hands into hills and covered them with a hoe but later a variety of hand planters were developed which eliminated both hand dropping and covering with the hoe. A farmer could plant up to two acres a day with one of these hand planters and it proved adequate for most of our relatively small New York farms growing only a few acres of corn. The author can remember his dad planting a 30-acre field of corn with this method in 1938, when a wet spring prohibited planting with his two-row horse planter. The large corn and wheat farms in the Midwest drove the need for larger planters, drills and harvesters whereas our New York farmers with smaller farms often clung to older methods.

Hedrick cites a U.S. Department of Agriculture survey stating that one man using a mechanical corn sheller could shell the same amount of corn as 100 men without a corn sheller. The survey was made early in the 20th century so with 21st century combines efficiency has probably increased another hundredfold. The need for improved methods

Planting corn, circa 2009.

A horse on a treadmill powering belt-driven buzz saw cutting wood into stove firewood lengths, circa 1900. Courtesy of OHA Museum & Research Center.

for shelling corn became obvious to thousands of farmers and resulted in over 300 different manufacturers of corn shellers.

The invention of the grain reaper is considered a major milestone in the world's agriculture. Cyrus McCormick received a patent in 1834 for what is considered to be the first successful reaper, although Obed Hussey invented one about the same time that was also successful. Both of these machines only cut the grain so it was still necessary to bind it by hand. Initially the reaper was also used to cut hay, but soon there was a specialized mowing machine to cut hay with the reaper used for cutting small grains. Improvements continued until machines were produced that would also tie the grain into bundles. This saved tremendous amounts of labor, as it took several people to bind the grain that one reaper cut. Walter Wood of Hoosick Falls, New York began to manufacture a reaper in 1852 and by the 1890s was said to be the largest agricultural machinery manufacturer in the world. The D.M. Osborne Company of Auburn, New York was formed in 1858 and became one of the largest reaper and binder manu-

facturing companies. In the 1900s, the Osborne Company was sold to International Harvester.[2]

The Johnston Harvester Company of Batavia was also noted for its reapers. In 1879, 95% of all reapers manufactured used Johnston's inventions. Samuel Johnston invented a corn and bean planter in 1855 and also a bean harvester. In 1868, he organized the Johnston Harvester Company, which was sold to Massey-Harris in 1910.

Syracuse became the center for the manufacture of cutting knives for mowers and reapers in 1858 when William A. Sweet established Sweet Brothers & Co. This company evolved into a number of other companies over the years, eventually becoming George Barnes & Co., which produced almost three million cutting sections in 1874, enough to provide sections for 181,888 cutter bars.

The first practical threshing machine was patented in 1837.[3] It eliminated the laborious task of flailing

2   Hedrick, Ulysses Prentiss *A History of Agriculture in the State of New York* p. 297-301

3   USDA Publication, *History of American Agriculture, Farm Machinery and Technology*

47

clockwise from top: Cutting oats with a grain cradle in Delaware County, circa 1915. Courtesy of New York State Archives.

Farmer using a handful of grain stalks to tie a bundle of grain. This was standard practice when using a grain cradle and reaper, but was eliminated with the invention of the grain binder. Courtesy of the OHA Museum & Research Center.

Loading bundles of grain on a wagon to draw to the barn or take directly to a threshing machine, circa 1932. Courtesy of the Liverpool Public Library's Schuelke Collection.

Three-horse team pulling a grain binder, circa 1920. Note the shocks set up on the right. After drying for a few days, the bundles were pitched on a wagon and drawn to the barn for threshing at a later date. Courtesy of Windsor Abbott.

An Allis-Chalmers combine, harvesting grain in the late 1930s.

A self-propelled combine, circa 1954. Prior to this time, almost all combines were pulled by a tractor. Courtesy of George Fesko.

Self propelled combine harvesting grain in Onondaga County, circa 1995.

TABLE 1 – A Comparison of the Hours of Labor, the Number of Acres and the Farm Equipment Used to Produce 100 Bushels of Wheat in the United States

From USDA, Economic Research Service (ERS), information in *History of American Agriculture - Farm Machinery and Technology*

Acres of Land, Labor Hours and Equipment to Produce 100 bushels of Wheat			
Year	Acres	Hours	Equipment
1830	5	250-300	Walking plow, brush harrow, sickle, flail
1890	5	40-50	Plow, seeder, harrow, binder, thresher, wagon, horses
1955	4	6-12	Tractor, plow, harrow, drill, combine, truck
1975	3	3 ¾	Tractor, plow, harrow, drill, combine, truck (all of these are larger and more expensive)
1987	3	3	Tractor, plow, harrow or disc, drill, combine, truck (all of these still larger and more expensive)

Note: The equipment listed has been changed slightly by this book's author to correspond with the situation existing in New York at those dates. Acres & hours are the same.

the grain. There was no separation of the grain from the straw and chaff so winnowing was still necessary. Gradually improvements were made with a carrier to remove the straw and later screens and a fan were added to remove the chaff and small trash from the grain. The early threshers were powered by a sweep with one to several horses traveling in a circle powering the cylinder. Later treadmills with one or two horses also powered threshers.

The *Old Mohawk Turnpike* book, from the Charles B. Knox Gelatin Co. Inc. edition, credits George Westinghouse with perfecting the threshing machine in 1840. Quite likely he developed a dependable working machine but with the ingenuity of thousands of inventors nothing remains perfected long because someone will think of a way to improve it. Mr. Westinghouse also produced a variety of other agricultural equipment in a Schenectady shop that in 1921 became the international headquarters for General Electric.

After the Civil War, steam engines gradually came into use to power threshing machines. From 1852 to 1892, Wood, Taber & Morse produced steam engines in the Madison County village of Eaton. At first they were stationary; later they were placed on wheels and pulled by horses. In 1882, a transmission was added making them four-wheel driven. There were many steam engine companies including the Birdsall Company of Auburn.

Mechanization dramatically reduced the number of hours needed to produce farm crops as is shown

vividly in **Table 1**. In the 157 years between 1830 and 1987, the labor required to grow 100 bushels of wheat was reduced by 99%! The acreage required to produce the 100 bushels was reduced by less than half. This reduction in acres is the result of improved seedbed preparation, increased yielding varieties of wheat and improved fertilization. This data is national. If figures were available for New York, the reduction in acreage to produce 100 bushels of wheat would be even greater.

The larger farms with large fields in the mid-western states encouraged the development of larger machines than were commonly used on our smaller New York farms. Tractors and combines came into general use in these states prior to their use in New York. The table is reasonably accurate for New York, however, for both acres and hours to produce 100 bushels of wheat. The trend of increased productivity has continued since 1987 and if figures were available, both acres and hours would be less at the time of publication.

Table 2 shows similar changes with corn.

It is unfortunate that we do not have information as to the number of hours it took to produce 100 bushels of corn in the early 1800s before corn shellers were put into use. The hours shelling corn by hand would have increased the number of hours dramatically. Until the time of the combine, a bushel of corn normally was with the kernals of corn still on the cob.

Plowing with an ox, circa 1900.

Plowing on Long Island with an early farm tractor, circa 1920.

Fitting a field with a disc-harrow for planting, circa 2009.

**TABLE 2 – A Comparison of the Hours of Labor, the Number of Acres and the Equipment
Used to Produce 100 Bushels of Corn in the United States**

From USDA information in *History of American Agriculture – Farm Machinery and Technology*

Acres of Land, Labor Hours and Equipment to Produce 100 Bushels of Corn			
Year	Acres	Hours	Equipment
1850	2 ½	75-90	Walking plow, harrow, hand planting
1890	2 ½	35-40	Walking plow, harrow, planter
1945	2	10-14	Tractor, larger plow, larger harrow, larger planter and a cultivator, corn picker
1975	1-1/8	3-1/3	Tractor, plow, harrow, planter, combine (all larger and more expensive), herbicide applicator & truck
1987	1-1/8	2-3/4	Tractor, plow, harrow or disc, planter, herbicide applicator, combine, truck (all of these still larger and more expensive)

Note: The equipment listed has been changed slightly by this book's author to correspond with the situation existing in New York at listed times. Hours & acres remain the same.

Mechanization began impacting the dairy industry after the middle of the 19th century. Cheese factories started to appear in many of our rural communities giving the farmer the choice of either making cheese from his milk at home or taking it to the cheese factory. Farmers gradually chose the cheese factory and often decided to add additional cows to their herds since they were now relieved of the burden of cheese making.

The increased number of milk cows encouraged creative minds to devise a method superior to hand milking and authorities indicate that in the 1870s the first successful vacuum milker came into use in the United States. Ezra Cornell, a decade earlier, however, had reported three persons in England using the "Yankee cow-milker", successfully replacing the labor previously required of 20 persons. It was many years before the majority of dairy farms used milking machines, but the door was open for increased mechanization on the dairy farm. S. M. Babcock, the inventor of an accurate test of the butterfat in milk, wrote in the 1892 *National Dairyman,* that "milking machines would result in poorer quality of milk and lower the standards of dairy animals". Very often new machines were marginal in their improvements and additionally had to prove themselves over a period of time before they were accepted.

In the 1870s, the first wooden silos began to appear. These first silos were square, made of wood and usually attached directly to the dairy barn.

The country creamery came into existence during the last quarter of the 19th century. The creamery gave the farmer the option of making his butter at home or taking his cream to a creamery. The removal of the butter-making chore from a farm tended to tempt the farmer to add additional cows. The cream was skimmed from the top of pans for delivery to the creamery until into the 1890s, when cream separators came into wide use.

In the introduction of the 1860 U.S. Agricultural Census, the following statement was made: "Land is abundant and cheap, while labor is scarce and dear." Although land in New York was more expensive than in most of the other states, labor was dear and New York was the leader in agricultural mechanization. A table in that census lists the value of farming implements and machinery in use for New York at over twenty-nine million dollars. This value was almost seven million dollars greater than Pennsylvania, which was second.

The census also stated that in 1860 Rensselaer and Cayuga counties each produced upwards of $400,000 worth of agricultural implements. There were 23 manufacturers of shovels, spades, hoes and forks in New York that year. According to the census in 1831, an estimated 5,000 dozen shovels worth $35,000 were made annually in New York. In 1849, the D.G. Millard Factory of Clayville, New York, makers of scythe and cast steel forks and powered by water, made 13,000 dozen scythes annually.

left: Cows were milked like this in the early 1800s before barns were constructed. Courtesy of the LaFayette Historical Society.

below: Removing the milking machine from a cow, circa 1950. Cows were milked with milking machines similar to this from the time electricity came to the farm in the 1930s, until replaced by the milking parlor about 40 years later. Courtesy of Windsor Abbott.

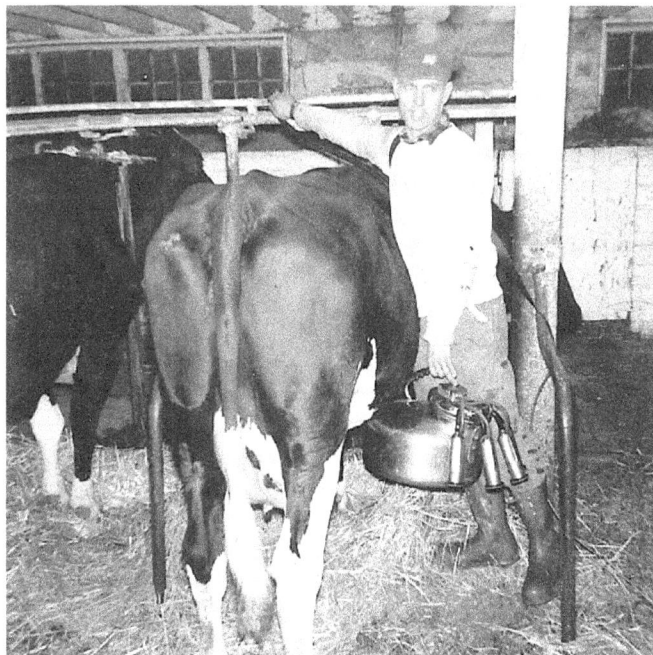

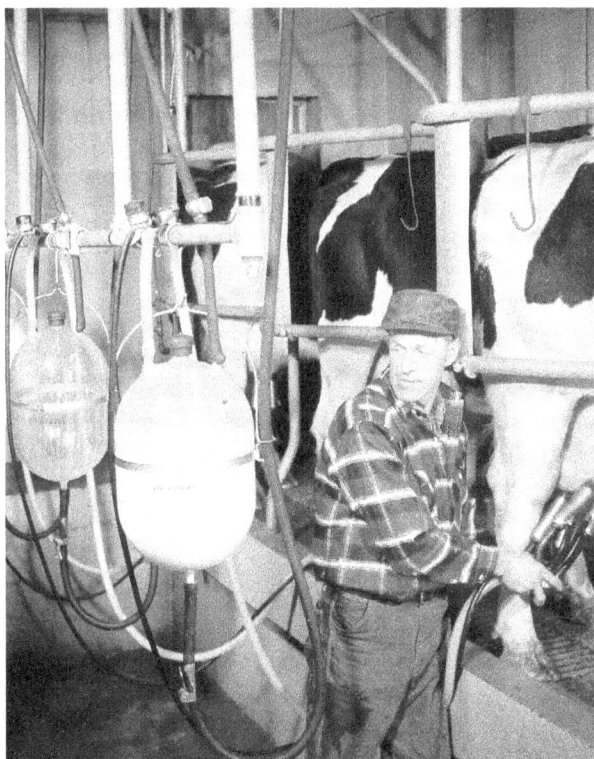

above: One of the first milking parlors, constructed in 1967. Courtesy of George Fesko.

right: A robot milking a cow at the Wolf Family Farm in Lyons in 2009. The cows are totally milked by the robot without the need for a person to be around. Each robot milks 50-55 cows with the cows coming into the robot and leaving of their own volition.

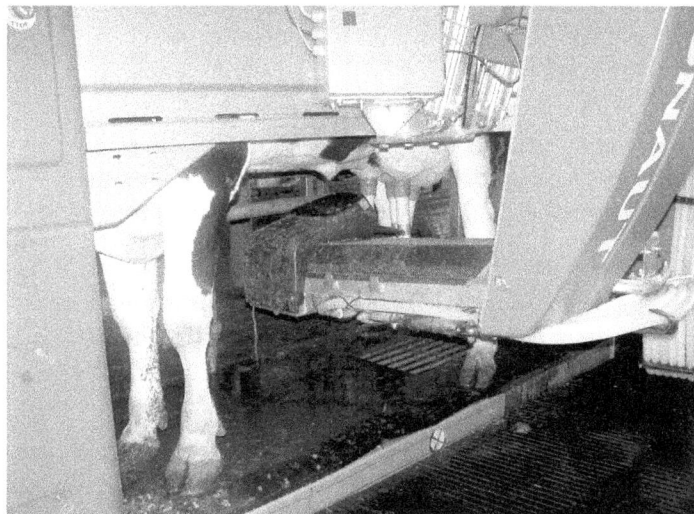

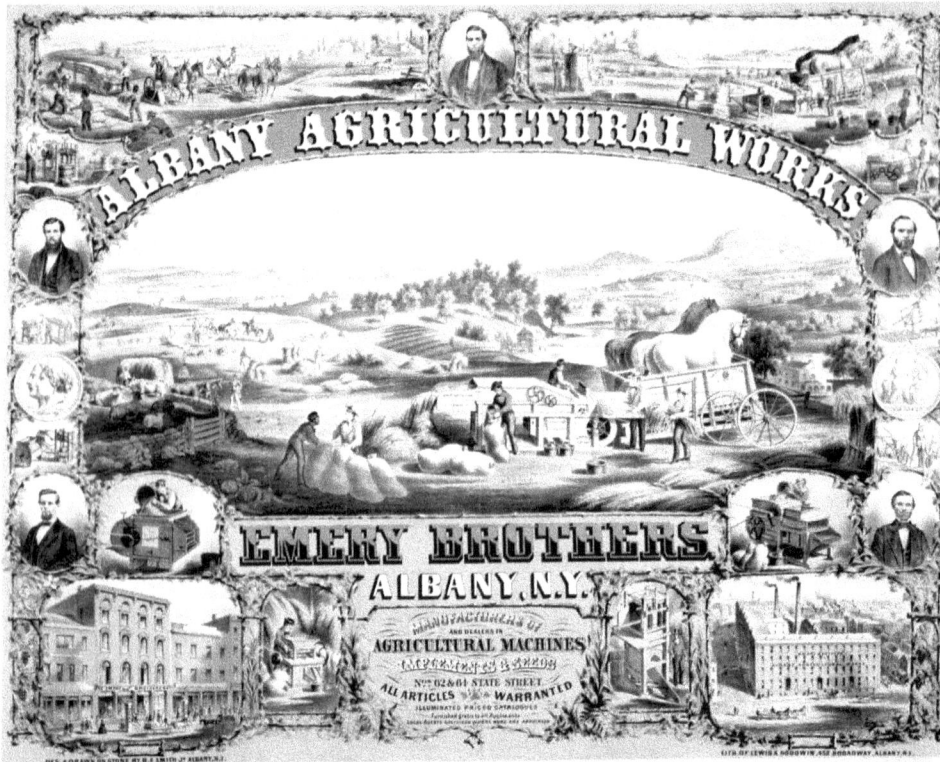

An advertisement by Albany Agricultural Works, 1865-67. Note the horse treadmill powering the threshing machine. Courtesy of the Albany Institute of History and Art, 1964.67.

change came slowly to the vast majority of farms.

The Civil War increased mechanization on New York farms due to the shortage of manpower caused by many able-bodied farmers off fighting in the war. A shortage of agricultural produce and higher farm produce prices encouraged increased efficiency and the purchase of more efficient machinery. This increase in agricultural mechanization also created the need for additional capital compared to the relatively small outlays needed to purchase farm machinery a few years earlier.

Ezra Cornell, retiring president of the New York State Agricultural Society, in his address at its annual meeting on February 12, 1863, provides us with a clear perspective of the changes taking place in agricultural mechanization at that time. Following are some excerpts.

> *"More than 100,000 laborers have been drawn from the tillage of the soil (in New York) to the destructive pursuits of war." "It was customary to expend from four to six days of manual labor in hoeing an acre of corn. Now, one day, with a horse-hoe, or an improved cultivator is adequate to the tillage of an acre." "Aged and infirm members of the household can cut with the machine (mower) as much grass in one day as ten of the most stalwart men could cut with the scythe in the same time."*

New York farm mechanization continued after 1875, but prices of agricultural crops decreased after the end of the Civil War and the number of New York farms decreased by about 15,000 between 1880 and 1890. Farmers observed the increased efficiency on neighbors' farms, where new labor-saving equipment was being used, but often could not afford to upgrade because of the small size of their farms. Mechanization continued slowly through

The 1860 census' introduction, apparently written in 1865, provides interesting data regarding some manufacturers of mowers and reapers in New York. It states that Adriance, Platt & Co. made 2,500 "Manny" and 1,100 "Buckeye" reaping and mowing machines. S. M. Osborne & Co. of Auburn made 15,000 "Kirby" mower and reapers and that the Buffalo Agricultural Works made 7,000 of the same machine. R. L. Howard, also of Buffalo, made 20,000 Ketchum mowing-machines and 5,000 reapers and mowers combined. Walter A. Wood of Hoosick Falls made over 30,000 reaping and mowing machines, of which over 1,000 were sold in Europe each year. Manufacturers of agricultural machinery in New York were busy not only supplying machinery to the farmers of New York but also farmers in other states and countries.

Some scholars refer to the period from 1862 to 1875 as the first American Agricultural Revolution. This is characterized by a change on the farm from hand power to horsepower. Change had been building slowly for half a century, with progressive farmers adopting new labor saving machinery on their farms. These farms were the unusual ones, however, and the remainder of the 19th century and into the 20th century when consolidation of farms and the recurring shortage of labor increased the pressure for farm mechanization.

Agriculture changed dramatically during the last half of the 19th century and change continued even more rapidly in the 20th century. During the last 50 years of the 19th century, the number of farms in the United States increased from 1,449,073 to 5,739,657. The acreage of improved farmland in the US had more than tripled from 1850 to 1900, while the area of improved farmland in New York increased by one quarter.

In 1850, New York had the most farms of any state with 170,621, but by 1900 six other states had more farms. The population of New York increased from three million to over seven million, while the center of the United States agricultural production gradually moved westward.

During the 20th century, New York's agriculture gradually changed to meet both the needs of the State's rapidly growing population and competition from agriculture outside of the State. With a population increasing from seven million in 1900, to almost nineteen million in 2000, New York became an attractive market for agricultural producers, not only in the United States but also throughout the world. New York's population growth presented both opportunities and challenges for our farmers.

Farmers in proximity to growing population centers found a ready market for perishable products that were difficult to ship from distant points of production. Vegetables harvested for sale increased from a little over 100,000 acres in 1900 to over 200,000 in 1950. Whole milk sales increased, while the production of butter and cheese, more easily shipped long distances, decreased. The invention of refrigerated railroad cars in 1878 brought dairy products from the west that competed with New York dairy products and had a depressing effect upon the value of New York's grazing lands, resulting in numerous hilly pastures reverting back to woodland. The production of oats, commonly fed to horses, decreased by more than half as the number of horses on New York farms decreased from over 600,000 in 1900 to less than a quarter of that number by the middle of the century. The acreage of other grains also decreased with the exception of corn, which was used as a grain to feed New York's livestock and as silage for dairy cattle. The number of hogs grown in New York decreased dramatically as production increased in states with rich soils and lower production costs.

In 1900, almost every farm had several kinds of livestock and crops to provide income. This

Steam engine, threshing machine and water wagon heading to a farm to custom thresh grain, circa 1910. Courtesy of Marcellus Historical Society.

TABLE 3 – US Census Figures

Year	Horses (000s)	Tractors (000s)	% Farms with Tractors	Value of Machinery & Implements (000s)
1900	631			56,006
1910	595			83,644
1920	543	7	3.6	169,866
1930	326	40	23.6	173,606
1940	275	58	34.7	138,341
1950	138	119	65.6	NA
1969	46	114	91.5	639,728
1997	47*	84	93.2	1,906,163
2007	85*	100	88.4	3,546,137

* Horses and Ponies

diversification served as a form of insurance to prevent total loss of income because of a crop failure or low prices. With a variety of crops there was less chance of a drought affecting all of the crops because of different planting and harvest times. Any farmer who had farmed for more than a decade knew that the price of any crop varied dramatically with supply and demand, but almost never knew what the harvest price would be at the time he planted his crops. As specialized equipment for a specific crop came into use in the 20th century, the cost of that equipment required the planting of additional acres to make the equipment profitable causing a decrease in diversification. This trend existed but was barely noticeable until tractors, and the larger equipment designed for use with tractors, came into being in the 1920s. Specialization in production became commonplace after the 1940s.

During the 19th century, almost every farm had an orchard with a mixture of apple, cherry, prune, peach and pear trees. These trees were planted to provide fruit for the farm family with perhaps a little surplus for sale. These trees were planted on whatever ground was available regardless of its quality and the trees often received little care. In the 20th century, most farmers gradually eliminated their orchards and bought fruit from farmers specializing in fruit production. The number of fruit trees in the State dropped dramatically but the total yields did not change appreciably because the fruit was now being grown where the climate and land were superior, and with better orchard care.

Little commercial fertilizer was used for crop production prior to the 20th century. Land was fertile and cheap with little consideration given to its long-term production. The organic fertilizer from livestock production was returned to the land, but farmers often did not recognize the importance of this fertilizer in producing a good crop. In 1900, the total expenditures by New York farmers for commercial fertilizer was about 4 ½ million dollars. Fertilizer costs more than doubled by the 20's and surpassed 30 million dollars in the 60s. In 1997, commercial fertilizer costs were 93 million dollars. Farmers found that to obtain the greatest profit it was necessary to feed the soil and increase yields.

Census figures in **Table 3** showing the number of horses, the number of tractors, percent of farms with tractors and the total value of machinery and

An early tractor for the garden farmer, circa 1920.

equipment on our New York farms, demonstrate the changing face of agriculture.

The number of horses on New York farms gradually decreased after 1900. Although the first gasoline powered tractor appeared in 1901, relatively few horses were replaced by tractors until after 1920. Recent census figures of horse numbers show an increase because ponies have been included. The main reason horse numbers have increased since 1940, however, is their use for recreation. There are many New York farms that provide boarding stables for suburban horse owners. There have also been an increased number of farms using horses for farm power during the past 25 years because of a renewed interest in traditional agricultural methods.

The percentage of farms with tractors has changed very little during the last 40 years with about 90% of farms with tractors. The census criteria defining a farm for the 2007 Census of Agriculture is: *a farm is any place from which $1,000 or more of agricultural products were produced or sold, during the Census year.* The definition of a farm has sometimes varied from one census year to the next, decreasing or increasing the number of farms in the State depending upon the change. Farms with sales of less than a few thousand dollars would often have no need for a tractor.

The first tractors tended to be large and cumbersome. The first successful light tractor was developed in 1926. The early tractors ran on steel lugs that provided traction and in the 1930s, the all-purpose rubber-tired tractor with complementary machinery came into use. By 1969, there was the equivalent of two tractors on each commercial farm in the State indicating that many farms had more than two tractors. In the 1940s, the trend was toward bigger tractors that could pull larger equipment. The early tractors were generally of 15 to 25 horsepower and might pull a two-bottom plow. The size has gradually increased with some of today's larger tractors over 300 horsepower and capable of pulling a twelve-bottom plow with wider and deeper furrows. Some of the smaller farms do not have need for a tractor while others still find tractors in the 25 horsepower range effective for their needs. Large commercial farms have a number of tractors of varying size for specific needs.

The value of machinery and implements on New York farms provides a clear picture of the mechanization of our farms. In 1900, the value of machinery and implements on each farm averaged a little over $200. The power on the farm came from the farmer's manual labor and horsepower with only simple tools required. The decrease in the value of the dollar over the past century contributes only to a small part of the increase with the average value of equipment for each farm now over $97,000.

At the beginning of the 20th century, for about 20 years, the steam engine had been used for powering threshing machines and a few other farm jobs. Only an occasional large farmer could afford a steam engine and a threshing machine, so when they were purchased the owner did custom threshing for a large number of farmers in the community. This practice continued into the middle of the 20th century with the gasoline tractor replacing the steam engine in the 1920s.

The advent of the small combine enabled many farmers to own their own combine and the threshing machine gradually disappeared. Combines were not common until 1945 when over 5,000 are listed in the census, then doubling to over 10,000 in 1950. The advent of the pick-up baler eliminated the harvest of loose hay. Some 90,000 balers were listed in the 1950 census. Many small dairy farms that used these small combines and pick-up balers went out of business in the last half of the 20th century and the number of combines has dropped to a little over 3,000 and the number of balers to 20,000. Combines and balers were much larger at the end of the century than in 1950.

World War I and World War II each fueled major changes on the farm. Both the need for increased food supplies and the shortage of labor put pressure on the farmer to produce more food with less labor. Higher prices for farm products, common in times of war, also encouraged greater production and the incorporation of labor saving tools and methods. When the wars ended, agriculture found itself in a situation of over-production and low prices creating a situation where marginal farmers went out of business and successful farmers mechanized and became more productive in order to survive with the prevailing lower prices. This trend is shown quite clearly in **Table 4**.

During the period of 1910 to 1920, encompassing World War I, the number of farms dropped by 10%

**TABLE 4 – Number and Size
of New York Farms**
US Census

Year	Number (000s)	Average Size in Acres
1900	226	100
1910	215	102
1920	193	107
1930	159	112
1940	153	112
1950	124	128
1960	82	164
1969	51	196
1978	43	220
1987	37	223
1997	31	228
2007	36	197

Note: Census criteria for a farm has changed several times but comparisons are valid

and the size of the farms in acres increased by 5%. During the period of 1940 to 1950, encompassing World War II, the number of farms dropped by 19% and the size of the farms in acres increased by 28%. The changes in the number of farms and the size of the farms are even more dramatic during the decade after each of the wars. Both decades following the wars were very difficult for farmers and brought significant change.

United States government food policy has encouraged low prices for the consumer. This policy has required New York farmers to remain price competitive with the prices of food products grown in the other states and throughout the world. Farmers have been forced to become as efficient as possible and in doing so the survivors have had to specialize their production and increase the size of their operation in order to afford the most cost-effective machinery available. (Changes in the average investment for machinery are shown in **Table 3**)

The change from ten-gallon cans to bulk milk tanks in the 1950s and 60s encouraged the rapid decline in the number of our dairy farms. Many of the farms with small dairies chose to discontinue their dairy rather than go to the expense of installing a bulk milk tank and increasing the size of their herd

to afford it. It was similar to the situation existing on the dairy farm 75 years earlier when a farmer had to build an icehouse if he wanted to ship his milk to a creamery. The bulk milk tank was only one of the expenses faced by the dairy farmer, who also needed a gutter cleaner, a field chopper and other labor saving equipment to efficiently produce milk. Census information shows the number of farms with dairy cows dropped from 43,582 in 1960 to 24,775 in 1969 and down to 5,683 in 2006.

The changes in dairying were mirrored in other areas of agricultural production. Farms with livestock, poultry, fruit and vegetables all began to specialize, increase in size or discontinue farming. Many farmers chose to continue to live on the farm while taking a job off the farm to provide income and still do a little farming on the side, either for the joy of it or to provide some additional income. This situation exists in the 21st century with many of the operators of the remaining 36,000 farms working off the farm to provide supplementary income for the family. The 2007 Census shows that over one-half of the farms have sales of less than $10,000.

Production per acre and production from each livestock unit has been increasing dramatically over the last century. Research has resulted in improved livestock, seeds, fertilizer and feeds, all increasing yields and often increasing the nutritional value of the products produced. Herbicides and pesticides have diminished many of the detrimental effects of weeds and insects although there has been a trend in the last 20 years toward either natural or organic production.

During the last half of the 20th century there was a strong movement toward conservation of soils with minimum tillage and no-till agriculture. There has also been a trend to minimize the amount of chemicals used on the land. The book, *Silent Spring*, by Rachael Carson, published in 1962, is widely credited with launching the environmental movement. It inspired public concerns of the use of pesticides and significantly increased concern regarding environmental pollution. The "cheap and plentiful supply of land" attitude of the 19th and early 20th century has been gradually changing, to one of treating both our land and our environment as valuable and indispensable assets, the quality of which must be preserved for future generations.

Chapter VII
Farm Numbers, Sizes and Prices

Initially, when New York was developing as a colony and later as a state, the land tracts were very large. They gradually diminished in size as parcels were sold to individual families who provided the greater part of the labor force on the farm. The amount of land that could be farmed by a family, using a team of oxen and later horses, was usually less than 100 acres even with half of that left in woodland and used for pasture. A few of the more successful farmers and landowners, whose wealth came from sources other than farming, owned farms of several hundred acres but they were in the minority. It wasn't until the arrival of the gas-powered tractor on the farm, accompanied by labor saving machinery, that farm size increased and the number of farms decreased.

The New York State Agricultural Census of 1875 provides us with a snapshot of the average farm by showing the average production of several key items on New York's 241,839 farms, as shown in **Table 5**. The farm family consumed a large part of the items it produced, with the exception of wheat and milk.

Farming with oxen or a team of horses could be accomplished on steep hillsides. Fields did not need to be large for them to operate effectively. When the tractor replaced the horse, fields needed to be larger and steep hills were not safe for the use of motorized equipment, so many fields that had been tilled were designated for use as permanent pasture or allowed to grow back to trees.

Even though farms became larger with the use of tractors and other farm machinery the amount of land used for agriculture decreased, as much of the land in the State was not suitable for use with mechanized equipment. Another factor that caused farmland to disappear, was the increase in New York State's population, from only 340,000 in 1790, to over 7 million in 1900, and to about 19 million at the present time. A great amount of farmland was turned into housing and the infrastructure needed to accommodate this increase in population. Along with this, we gradually moved from an agrarian state to a manufacturing state requiring thousands of additional acres for non-farm purposes. **Table 6**

TABLE 5 – The Average Production per Farm in 1875
New York Census

Item	Amount	Item	Amount
Oats	157 bushels	Wheat	43 bushels
Potatoes	151.5 bushels	Apples	95.6 bushels
Cider	2.2 barrels	Butter	446 pounds
Cheese	32.2 pounds	Milk sold	171.7 gallons
Wool	30.5 pounds	Pork made on farm	501 pounds

TABLE 6 – Number of Farms and Size in Acres
(in thousands) US Census

Year	Total Number	Less than 10	10 to 175	More than 175
1900	226	17	178	31
1920	193	14	147	32
1940	153	13	111	29
1964	66	4	36	26
1978	43	2	21	19
1997	32	2	17	13
2007	36	3	22	11

TABLE 7 – Number of Farms with Value of Sales
(in thousands) US Census

Year	Less than $2,500	$2,500 to $25,000	$25,000 to $100,000	Over $100,000
1969	18	34*		1
1978	10	15	14	4
1987	9	12	9	7
1997	8	11	6	7
2007	13	12	5	7

*The two groups are combined in the census
Note: Numbers are rounded to the nearest 1,000

Plowing and harrowing on the Arbuckle Estate in the Wallkill Valley near New Paltz in Ulster County, circa 1920. Wealthy men sometimes had large estates employing a number of men and producing many times more produce than the average farmer. Courtesy of New York State Archives.

shows the number of farms by size, at several time periods during the 20th century and also in 2007.

The number of farms with less than ten acres has remained between 5 and 10% of the total number of farms throughout the 20th century. The number of farms with over 175 acres has increased from about 15% to 30%. This increase is not surprising, as today's dairy, poultry and cash crop farms need additional acres to remain competitive and profitable.

Farm size, in acres, has varied significantly from one county to the next. The lay of the land, the predominant type of agriculture, the overall population and the climate all have a bearing on the size of farms. In 1860, Essex County had the largest farms with an average size of 199 acres, followed by Hamilton with 167 and Delaware with 147. In 1950, Delaware County was first with 195 acres, followed by Lewis with 189 and Livingston with 178. In the 2007 Census, Jefferson is first with 296 acres, followed by Wyoming with 287 and Livingston with 281. In all census periods, the farms with the smallest acreage are those in the greater New York City area. (Census information for all of the counties is listed in the appendix)

Hamilton's average size farm shrunk from 167 acres in 1860 to 23 in 2007 because of its location in the Adirondack Forest Preserve. The counties with the largest farms, by acreage, tend to be counties with large dairy farms where most of the feed for the cows is produced on the farm.

Another measure of a farm's size is the value of the products sold from the farm. In recent years the number of farms are also listed in the census by the value of the sales from the farm. **Table 7** provides this information from census information available since 1969.

Often a farm that is small in acres has a large value of sales. This is often the case with farms that have greenhouses, nurseries, vegetables, fruit or a concentrated livestock operation. You will note that there has been a large increase in the number of farms with sales of less than $2,500 between 1997 and 2007. The increase is due, to a great extent, to the increased number of people desiring natural or organic foods. Many of them are growing farm products for their own family's consumption with a surplus for sale.

TABLE 8 – Number of New York Farms, Average Size and Value
US Census Data

Year	Number of Farms (000's)	Average Farm Size in Acres	Total Value of Farms (000,000s)	Average Value of each Farm
1850	171	112	555	3,250
1880	241	99	1,056	4,381
1900	226	100	888	3,917
1920	193	107	1,425	7,376
1940	153	112	947	6,180
1969	52	196	2,772	53,399
1987	38	223	8,263	218,934
2007	36	197	16,322	449,010

Farm values have varied considerably from the time New York became a state. Variations in the weather, creating either bountiful crops or famine, have caused farm values to decrease or increase. Prior to improved transportation provided by canals, railroads and highways, the size of a crop had little effect upon prices more than 100 miles away. Today, a crop failure in China, Russia or South America has repercussions around the globe. The value of a farm, based upon its production capabilities, is dependent upon the profit that can be made on the farm products produced. If prices are high, especially over a period of several years, land values increase. If prices are low for several years, land values decrease.

Table 8 illustrates the changes in the value of farms at several time periods between 1850 and 2007.

There are several striking comparisons shown in the table. The first is the decrease in the number of farms, especially the 65% decrease during the 29-year period from 1940 to 1969. Approximately 2 out of 3 farms ceased to exist during that period. During the same period the average farm size increased by 75%. It is evident that farmers continuing to farm were purchasing many of the farms going out of business.

The figures in the table that are most shocking are the significant decrease in the value of farms between 1880 and 1900, and also between 1920 and 1940. In the 20-year-period between 1880 and 1900,

Plainville Turkey Farm headquarters in Onondaga County, with turkey processing plant in the center of picture, circa 2005. This farm raised over 500,000 turkeys annually.

Hourigan Farms, a large dairy farm in Onondaga County, 2008.

the total value of farms in New York decreased by 16% and in the 20-year-period between 1920 and 1940 the value of New York farms decreased by 33%. Most of us have lived in a time when prices have increased because of inflation, so it is an awakening to note those significant decreases. If we had data to show the change each year during those periods, we would likely see changes to the upside as well as the downside, but with the overall trend downward.

The nearly 100% increase in the value of farms between 1850 and 1880 was due to inflation during the Civil War period and also because farmers throughout the State were improving their farms. New York agriculture was still in a period of growth during those years, with farmers building new barns and homes as well as increasing the acreage of improved land. During the period from 1940 to 2007, the total value of New York farms increased by 17 times, while during the same period the consumer price index increased by 14.5 times and the acreage of farmland decreased by 76%. There were many building improvements on the farms during this period, but farmland increased in value faster than the increase in the consumer price index. In 1940, the value of New York farms averaged $55 per acre, while in 2007 the average value of farms per acre had increased by 23 times to $1,284 an acre. When dealing with averages, we need to realize that the increased population of the State has placed significant pressure on land prices near population centers. The 2007 Census

shows the value of farms per acre, including buildings, varied from $1,346 in St. Lawrence County to $98,997 in Nassau County.

The U.S. General Price Level from 1820 to 1914, indexed to the Consumer Price Index, varied between 6 and 20 with the lowest figure in the 1890s and the highest during the Civil War. At first glance, a change from 6 to 20 seems minimal until we recognize that it represents more than a 300% increase and a drop from 20 to 6 would be a decrease of 70%.

Success in farming depends on many things, including size of farm, quality of the soil, weather, management capabilities and financial resources. Another important factor is the time a farmer is born. The opportunity of success for a beginning farmer, prior to the great depression of 1929 to 1939, when farms were expensive and were subsequently followed by low crop prices, was far less than

TABLE 9 – Consumer Price Index by US Dept of Labor on January 1

Year	Index	Year	Index
1914	10	1970	37.8
1920	19.3	1980	77.8
1930	17.1	1982-84	100
1940	13.9	1990	127.4
1950	23.5	2000	168.8
1960	29.3	2009	211.1

for one starting farming in 1939, when farms were cheap, prior to the boom years of 1940 to 1956.

The Consumer Price Index Data for the years from 1914 to 2009, with the decades in between, are listed in **Table 9**.

The table shows prices almost doubling during and shortly after the two World Wars as well as the large decrease in prices during the Great Depression. The consumer price index in 2009 is 21 times larger than it was in 1914. Almost any agricultural commodity would have had even greater fluctuations during the various time periods than the Consumer Price Index, because of the great variation in the supply of the various farm commodities.

A History of Agriculture, Economic Cycles by the Economic Research Service of the USDA lists the economic cycles from 1776 to 2000. During that period of 224 years there were over 30 time periods of significant price increases and decreases, as shown in **Table 10**. A person farming for a typical 40-year-period experienced at least two cycles of price increases and decreases. These economic cycles, coupled with the normal erratic weather patterns that produce either bumper crops or crop failure, make surviving on the farm a real challenge.

Although **Table 10** shows the major economic cycles, often there were shorter cycles within these cycles. Agricultural cycles normally do not mesh closely with economic cycles because of variations in weather, long-term storage of grains and price peaks and valleys of the various agricultural crops arriving at different times. Just surviving until there was no light at the end of the tunnel, is what happened to many thousands of New York farmers. The drop in the number of farms from over 226,000 in 1900 to the current 36,000 is only part of the story. Many farms changed ownership several times, with one

TABLE 10 – List of Economic Cycles
US Department of Agriculture

Years	Prosperity	Years	Recessionary
1776-83	Revolutionary War Boom	1784-88	Postwar depression
		1807-09	Embargo depression
1810-14	Uneven wartime prosperity		
1815-19	Speculative boom	1819-22	Panic and depression
1823-33	Gradual recovery	1833-34	Banking recession
1834-37	Speculative boom in land, banking and transportation ends in panic of 1837	1838-43	Depression
1844-56	Recovery and business expansion	1857-60	Panic of 1857 and recovery
1861-65	Civil War prosperity & inflation	1866-67	Postwar recession
1868-73	Railroad boom	1873-78	Depression & deflation
1879-93	Business expansion	1893-94	Depression
1895-96	Return of prosperity	1907-08	Panic of 1907
1909-14	Prosperity		
1914-18	War boom	1920-21	Sharp postwar recession
1922-29	Speculative boom	1929-39	Great depression
1939-45	World War II	1950-56	Korean War and readjustment
1946-70	Postwar boom	1957-58	Recession
1958-70	Extended business expansion	1970-80	Slowing economic growth and inflation
1983-90	Business expansion	1981-82	Recession
		1990-91	Recession
1991-00	Business expansion		

TABLE 11 – USDA Economic Research Showing Total Dollars Spent for Food in the US and Percent of Disposable Income Spent for Food at Home & away from Home

Year	$ spent for food (billions)	Percent spent for food at home	Percent spent for food away from home	Total percent spent for food
1929	19.5	20.3	3.1	23.4
1936	14.7	18.8	3.0	21.8
1942	22.1	15.6	3.0	18.6
1947	40.2	19.2	4.3	23.5
1960	64.0	14.1	3.4	17.5
1972	117.3	9.9	3.6	13.5
1984	345.8	7.7	4.2	11.9
1996	593.8	6.3	4.1	10.4
2007	997.3	5.7	4.1	9.8

owner after another deciding that farming was not the correct choice for him. You can also observe from **Table 10**, that the opportunity for success can vary with when a person starts farming.

The Economic Research Service of the USDA also provides interesting information concerning the portion of the consumers disposable income spent for food from 1929 to 2007. **Table 11** gives several years of information to show the trends.

The number of dollars spent for food has increased by over 50 times since 1929. Significant portions of the increased cost are for added value, through greater preparation prior to the consumer's purchase, additional transportation costs because of production coming from greater distances (including increased imports from foreign countries), and a general increase in all costs due to inflation. (**Table 9** shows that the Consumer Price Index increased approximately12 fold during the same period of time.)

From 1929 to 2007 the percent of the consumers disposable income spent for food decreased from 23.4 to 9.8, a decline of 58%. The increased productivity and efficiency of agriculture during this 78-year period are almost beyond belief. In 1929, no one could have possibly foreseen the magnitude of the changes that have transpired.

Another factor that makes the change in the percent of disposable income spent for food more dramatic, is that the percent spent for home use has decreased from 20.3 to 5.7, a decline of 72%. You will note from the table that the percent of disposable income spent for food, eaten away from home, has been fairly steady with an increase from 3.1 to 4.1 percent. The dollars spent for food consumed away from home are now 42% of total consumer food costs compared to a little over 13% in 1929. Food eaten away from home has increased value-added costs, making the percent of actual food ingredient costs smaller than for food consumed at home.

The farmer's share of the food dollar has gradually decreased from 47% in 1952, to 38% in 1972, to 23% in 1997 and down to 20% in 2007. Through improved seeds, increased fertilization, research, mechanization and greater efficiency, farmers have been able to produce their products at lower and lower costs. This has provided billions of dollars for consumers to use for improved homes, better healthcare, more leisure activities and an overall higher standard of living than would otherwise have been possible.

Chapter VIII
Hay, Pasture and Grains

When we think of the early settlers cutting the forests for crops, we tend to think of the land being turned into the production of small grains and row crops. The first cleared acres were used in this manner while the livestock, especially the cattle and hogs, searched for food in the forest. As more land was cleared of trees and the number of livestock increased, much of the land that had been cleared was turned into pasture or utilized for the production of hay. In addition, thousands of acres of woodland were pastured well into the 20th century, but the acreage gradually decreased as farmers realized greater production with improved pasture and found that woodland grazing was detrimental to future woodland growth. As late as 1855, only 14.5% of New York's farmland was used for the production of crops. By the early 1900s, the percentage of farmland utilized for crops other than pasture reached the low 40s. The first

figures we have for the amount of woodland utilized for pasture is a little over 2 million acres in 1917. The acreage of woodland pastured has gradually diminished since that time to 165,855 acres in 2007.

When the forests were cleared, there was a large acreage of land that was not ideally suited for crop production because it was either wet or hilly. Much of this land was seeded for pasture and remained permanently in pasture. The 1855 census lists almost 5 million acres of permanent pasture and the acreage increased to a little over 7 million in 1930. Permanent pasture acreage decreased gradually to 714,000 acres in 2007. **Table 12** provides a view of the changes.

Hay and pasture have utilized more of New York's farmland than any other crops since forests covered New York. The percent of farmland used for hay and pasture is shown in **Table 13**.

TABLE 12 – Hay and Pasture Acreages
(in thousands) NY State and US Censuses

Year	Hay & Hay Forage	Permanent Pasture	Woodland Pastured	Cropland Pastured
1855	3,384	4,984	NA	NA
1875	4,797	5,976	NA	NA
1917	4,304	4,682	2,138	NA
1950	3,196	6,066	1,362	NA
1997	2,073	473	226	633
2007	1,963	715	166	280

NA – No information available

Cows enjoying pasture on the Dennis Farm in 2008 at Pompey in Onondaga County.

TABLE 13 – Percent of New York's Farm Acres Used for Hay and Pasture

Year	NY Farm Acreage (000s)	Hay & Pasture Acreage (000s)	% of Farm Acreage
1855	19,119	8,368	43.8
1875	25,659	10,763	41.9
1917	19,092	11,124	58.3
1950	16,017	9,262	57.8
1978	9,461	4,237	44.8
1997	7,254	3,179	43.8
2007	7,175	2,958	41.2

Hay and pasture were important since they provided the bulk of the food required, from the early 1800s into the 1900s, for our large dairy industry, as well as our sheep, hogs and poultry. They served a similar purpose to our gasoline and diesel fuel of today by being the "fuel" for horses and oxen that provided the power on our farms and transportation on our highways. In the early 1900s, with the increased use of motorized vehicles, there was an agricultural depression in northern New York when it lost its market for hay in the greater New York City area. For many years, hay had been shipped by train to New York City to feed the thousands of horses in the area transporting people and delivering merchandise.

Traditionally, pastureland on farms has been the land that is more difficult to work because of hills and wet areas. The best land was used to produce annual crops with the poorer quality land serving as permanent pasture. The farms without sufficient hilly and wet areas for pasture rotated their pastureland similar to their cropland. One of the first projects in the spring, for the farmer with permanent pasture, was to repair the fences damaged by heavy snows during the winter. Soon after the arrival of electricity on the farm, electric fences were invented which made the fencing of temporary pastures more convenient.

Until the mid-1900s, pasture was important for both poultry and farrowing hogs. This pasture was planted especially for their use and located on ground that would later be plowed for corn as the waste from the animals provided excellent nutrients for the corn.

For more than a century, except during the winter, livestock received the majority of their feed by grazing on pasture supplemented with some ground grain. With the advent of the field chopper in the last half of the 20th century, pasture for milking cows was eliminated on many dairy farms and fresh green

Horse powered sweep press in Onondaga County circa 1895. Hay was pitched into the top of the baler and then a horse-powered screw compressed the hay into 200 pound bales. Courtesy of the DeWitt Historical Society.

Chopping high moisture hay for silage at Twin Birch Farms in Onondaga County in 2008.

Unloading and packing high moisture chopped hay in a bunker silo at Twin Birch Farms in Onondaga County in 2008.

chop, silage, hay and grain were fed in the barn all year. Replacement dairy heifers still continued to be fed on pasture. Recently, because of a consumer demand for organic milk, some farmers are again using pasture for their milking cows. Most pastures today are rotational grazed, which increases the quality and quantity of the forage.

Around 1900, Onondaga County became the leading county in the production of alfalfa. The high lime soils in the limestone belt of the County produced a large crop of high quality alfalfa. A strong market developed for the alfalfa with large quantities being shipped out of the County. In 1900, Onondaga County produced two-thirds of the alfalfa grown in New York State. Alfalfa production increased rapidly in both the County and the State during the next 30 years. The Onondaga County Alfalfa Growers Association had 100 members in 1916 shipping 4,000 tons of alfalfa. The average price was $13.40 a ton, about $2.40 more per ton than other hay. Maplehurst Farm, located between Fayetteville and Manlius, produced over 300 tons a year with three cuttings of the crop. Some farmers combined various clovers with timothy for hay but

alfalfa provided a greater yield. Alfalfa is still the first choice of most dairy farmers for both dry hay and high moisture hay.

New York's climate requires that approximately one-half of the feed for cattle and horses, which is largely hay, is harvested in the summer to serve as food for them during the late fall, winter and early spring months when pasture is not available. The method of cutting hay has gradually evolved from sickle, to scythe, to mowing machine and to huge self-propelled mower-conditioners used today. Hay was harvested in its bulky, loose form for thousands of years, until the pick-up baler came into use in the middle 1900s. Today, pick-up balers are still used to harvest dry hay and some farmers are using pick-up balers for high moisture hay that wrap the hay in airtight plastic bags, sealing the nutrients inside the bale until the bale is opened, months later, for feeding livestock. The physical labor associated with harvesting dry hay has largely disappeared as farmers are currently using balers that form large round bales, handled mechanically with tractors.

Dry hay is still baled for heifers and horses with a large existing market for producers of hay. The

2007 Census indicates hay production of 4,941,812 tons, including green chop and high moisture hay converted to a dry matter basis averaging 2.5 tons per acre. This compares to 4,783,857 tons in 1875 yielding 1.1 tons per acre. Hay acreage (**Table 12**) is less than half of that in 1875, but provides a larger total yield. The hay produced today is of far better quality because of earlier cutting and the development of superior varieties. Unless a person has had the opportunity to harvest loose hay, it is hard to imagine the amount of work that went into producing almost 5 million tons of loose hay each year. In the 1875 era, with over 240,000 farms, there would have been over half a million New York farm people involved in their farms' haying operation.

Currently, the most common method of harvesting hay for dairy cattle is with field choppers. The chopped, high moisture hay is transported from the field to bunker or upright silos in large trucks or wagons. Man no longer physically handles the hay, but instead operates the machinery that cuts, harvests and stores the hay. Hay harvesting methods and machinery continually change as hay remains a major crop for New York's livestock.

GRAINS

From the time of the first New York settlements by Europeans, grains have been priority crops for New York farmers. Corn was not native to Europe and though it was a mainstay in the diet of Native Americans, the colonists preferred to use wheat for their baked goods and relegated corn to their livestock, using it for themselves only when wheat was not available. Oats were the important grain for feeding horses as well as poultry and other livestock. Corn, barley and rye were used in the production of alcoholic beverages in addition to their use as livestock feeds. By the time New York became a state, the farmers were growing wheat, oats, corn, barley, rye and buckwheat.

We do not have information as to the quantities of each grain grown until New York's 1845 Census was published. It showed over a million acres of both wheat and oats grown, with total yields of over 13 million bushels of wheat and over 26 million bushels of oats. A bushel of wheat weighs 60 pounds and a bushel of oats 32 pounds resulting in a similar yield by weight for each crop. Barley and the other grains totaled a little over one million acres.

Cutting corn with a corn binder, circa 1931. Courtesy of the Liverpool Library's Schuelke Collection.

TABLE 14 – Acres and Yields of Wheat and Oats
NY and US Census

Year	Wheat Acreage (000s)	Wheat Yield per acre in bushels	Oat Acreage (000s)	Oat Yield per acre in bushels
1845	1,014	13	1,027	26
1865	512*	12	1,110	17
1900	558	19	1,330	31
1950	385	27	564	29
1987	86	42	163	58
2007	85	53	61	58

*One fifth was spring wheat yielding 8 bushels per acre

The midge, a small fly-like insect, and the Hessian fly invaded the wheat fields of Eastern New York in the early 1800s and gradually moved westward through the middle of the century destroying many fields of wheat. Alternative wheat varieties more resistant to the midge were planted but were less productive. A black stem rust had earlier invaded the wheat fields of New England and Long Island ruining wheat fields. As settlers moved westward in New York they escaped these problems for a time but their wheat fields eventually were infected. These infestations, coupled with increased wheat production in the Midwest, decreased the acreage of wheat grown in New York in 1865 by almost half.[1]

George Geddes, a prominent New York farmer, summed up the situation in an Agricultural Survey of Onondaga County, printed in the 1859 *Transactions of the New York State Agricultural Society*. He wrote, "The great and overshadowing obstacle to profits from farming is the ever present armies of insects destructive to every crop we raise."

Once the Erie Canal opened in 1825, wheat could be shipped economically from Western New York to Albany and down the Hudson River to worldwide markets. Much of the wheat grown in Western New York made a stop in Rochester, which became known as the "flour city" because of the quantity of wheat milled in that city for flour. Rochester was just a hamlet prior to the Erie Canal but waterpower, coupled with large quantities of wheat, turned it into a major city. The counties in proximity to Rochester are still the largest wheat producers in the State.

Genesee County's 23,298 acres make it number one, followed by Livingston and Monroe, both growing over 11,000 acres as reported in the 2007 Census.

The methods of growing and harvesting wheat and other small grains have undergone tremendous change since they were first grown in New York. **Table 1** in Chapter 6 shows the changes in the machinery and labor used to grow wheat at several different periods of time. The labor to produce an acre of wheat is approximately one hundredth of what it was 200 years ago and the yield per acre has more than tripled. Even with this increased efficiency and yield, the acres of wheat produced in New York continue to decrease because of lower costs of production in other parts of the world. **Table 14** provides the changes in acreage and yield of both wheat and oats in several census years.

Since horses consumed a large portion of the oats produced, the acreage of oats decreased proportionally to the number of horses. In **Table 14** you will notice a trend of increased yield over the 162 years represented, but not with an increase in each year. Yields of all the grains have increased because of improved varieties and fertilization, but when one compares yields on a year-to-year basis the yields may not follow the trend because of variable weather, varying from poor or excellent, seldom producing an average yield. Weather can also affect the number of acres planted in any year because wet fields, at the time of planting, require the farmer to select an alternative crop that can be planted later.

Straw, a by-product of small grains, has had many uses. Its main use on the farm has been as bedding

1 Hendricks, Ulysses Prentiss *A History of Agriculture in the State of New York* p. 332

Tractor mounted cornpicker, circa 1950.

for livestock, absorbing animal waste and then being applied to the land as a fertilizer. It also serves as mulch for strawberries and was commonly used to pack heads of cabbage for shipping and for winter storage. During the 1800s, large quantities of straw were drawn by the farmer and sold to paper mills for the manufacture of paper. There were a number of strawboard factories in New York in the last half of the 19th century.

Until the late 1800s and early 1900s, threshing of small grains was typically accomplished with a flail and winnowing board or a fanning mill. With the advent of the steam engine threshing of grain became more specialized, with one person owning a steam engine and a large threshing machine. This person traveled from farm to farm with his equipment, spending a day or part of a day threshing each farmer's grain. With the advent of the tractor the steam engine passed into oblivion, but this type of threshing continued into the 1940s when a small combine on each farm replaced the huge threshing machine. It took a crew of from 6 to 12 men to perform the various tasks in threshing so neighbors would help each other to provide sufficient labor. If the grain was in the mow of the barn it usually took three men to pass the bundles with forks to

the person feeding the thresher. A straw stack was often built in the barnyard with two men forming the stack as the straw was blown to the stack from the thresher. Two other men carried the threshed grain from the thresher to the granary across from the mow. If the grain came from shocks in the field it would take three or four wagons with a person loading each wagon and two men pitching the bundles on the wagon.

Threshing day was a big day on the farm. It was not only the culmination of a year's work growing the crop but it was also a day of shared hard work and joviality with the neighbors. The wife of the farmer, whose grain was being threshed, put on a big dinner for all of the threshers with meat, mashed potatoes, gravy, pie and all of the extras making certain that the hungry men were well filled. Many stories would be told at the table during the meal and then it was back to work for all the men.

New York farmers, from the days when New York was still a colony up until recent years, commonly grew barley and rye. Both grains were sold to distilleries and barley was used in the production of beer. Barley was a good feed for livestock and in the last half of the 20th century, when the US government imposed wheat quotas, barley was often grown on

TABLE 15 – Acres of Other Grains
(in thousands) NY and US Census

Year	Barley	Rye	Buckwheat	Soybeans
1845	192	317	255	
1865	243	233	18	
1950	25	16	73	7
1997	18*	8*	NA	100
2007	10	7	NA	200

*From New York agricultural statistics

New York farms in place of a portion of the normal wheat crop. Today rye is often grown for its straw, which is used for animal bedding and as mulch on strawberries. **Table 15** shows the acreage of barley, rye, buckwheat and soybeans grown in several of the census years.

Buckwheat flour was popular for use in baking into the middle of the 20th century but its use has since decreased. There is a mill in Penn Yann still processing buckwheat and selling buckwheat flour today. A century ago farmers often raised buckwheat. When there was a wet spring, farmers could plant buckwheat as late as early July and still obtain a crop.

Soybeans are a relatively new crop appearing for the first time in the 1930 census figures with 3,241 acres. The census information for that year indicates that they were grown mostly for soybean hay, grazed by cattle and for silage. By 1940, total soybean acreage increased to over 18,000 acres with not quite half of that harvested for beans. Since then soybean production gradually increased to about 26,000 acres in 1987 but by 1997 (see **Table 15**) soybean acreage had reached 100,000 acres and doubled during the next decade to 200,000 acres. The acres of soybeans now exceed all other grains with the exception of corn. Soybeans grow well on many of our New York soils and have offered a good return to the farmer. Even though the majority of the soybeans are marketed outside of the State because of the lack of a large soybean processing plant in the State, they have become a popular crop in recent years. Large quantities of soybean meal are now shipped into New York as feed for our livestock, which will increase the likelihood of new processing facilities here. The leading counties in the production of soybeans are Cayuga, Seneca and

Madison, each growing over 20,000 acres in 2007.

The sunflower is a grain crop that is beginning to be seen more often in New York. Sunflowers are grown for their seeds, which are a popular feed for wild birds or are crushed to produce sunflower oil. The 2007 census lists 357 acres, with a good portion of that acreage grown in Onondaga County. Sunflowers are harvested with a combine in a manner similar to the other small grains.

Corn has been harvested as a grain in New York for centuries. It is native to the Americas, probably originating in Central America where it resembled one of the small grains more than the large corn ear that we are familiar with today. Native Americans selected and planted corn with desirable traits for many years, developing several varieties that white settlers then grew in the 1800s and the early 1900s. Plant breeders have developed hybrids since the 1920s that have substantially increased corn yields.

Silos did not come into use until the latter part of the 19th century. Prior to that time and into the middle of the 20th century, dairy farmers without silos husked the corn and often chopped the dried corn stalks and stored them in the barn for the cows. The nutritional value of the corn stalks was relatively small without the ears. Chopping the green corn with ears attached provides a very good feed for cattle and was a procedure rapidly adopted by farmers in the early 20th century. Today corn silage is consumed by dairy cattle with a small portion used for beef cattle. Much of the corn grain is ground and fed to dairy cattle as well as other livestock and poultry.

The acreage and the tonnage of corn for silage were not tracked until the 1930 census. **Table 16** provides census figures for corn acreage and yields in several time periods. The increases in yields have been remarkable due to improved plant breeding, fertilization and weed control. With the acreage of corn doubling in the last 150 years and with the increased yields, the amount of corn produced has increased ten-fold. The 1845 census is the first

TABLE 16 – Acres (in 000s) and Yields of Corn
NY and US Census Data

Year	Total Acres	Grain Acres	Bu. / Acre	Silage Acres	Tons / Acre
1845	595	595	25		
1900	659	659	30	NA	NA
1930	558	111	38	340	8
1940	684	188	36	433	9
1950	624	163	47	439	9
1959	623	219	56	396	10
1978	1260	600	85	660	13
1997	1130	579	107	551	15
2007	1059	552	129	507	17

census listing corn acreage. At that time corn was referred to as "Indian corn" and there were 595,000 acres grown in New York.

Originally, corn was a very labor-intensive crop. After a field was plowed and fitted it was planted by hand, one hill at a time. It had to be cultivated and hoed to control the weeds. At harvest time it was cut by hand and the stalks were gathered together and placed in an upright position forming shocks. During the late fall or winter the shocks were dismantled and the ears husked for storage in the corncrib. The stalks were taken to the barn and fed to the cattle. As corn was needed during the year it was removed from the cob (shelled), for the poultry. When corn was fed to the hogs it was often left on the ear for the hogs to do their own shelling. The whole ear, including the cob, was usually ground for the cattle.

Dozens, even hundreds of different corn shellers were invented over the years. The earliest ones manually shelled one ear at a time with later mechanical machines shelling one ear after another, as the ears were fed into the sheller, while a hand-crank was turned to provide the power. In the early 20th century, a two-cycle motor powered larger corn shellers and in the 1930s a tractor powered still larger shellers.

With mechanization in the 1930's and 40's the mechanical corn picker, pulled by a tractor, saved many labor hours by eliminating the jobs of cutting, shocking and hand husking the corn. The combine, with a corn header, which came into use about 20 years later, saved even more labor by shelling the corn on its stalk in the field. The combine eliminated the need for the corncrib but added the need for a dryer and large storage bins for the shelled corn. My estimate of the labor requirements for growing an acre of corn on a large farm, from planting through harvest and shelling today, would be two hours compared to over 100 hours 150 years ago. The corn yield per acre today is 10 times greater than 100 years ago as a bushel of corn at that time was corn-on-the-cob and today it is shelled.

Another tremendous labor-saving invention, in the 40's, was herbicides to kill weeds. Sprays were developed that killed the weeds and did not hurt the corn. These sprays eliminated the need to hoe and cultivate the corn. More recently, corn has been genetically altered for resistance to improved herbicides that kill the weeds but do not harm the corn, and to provide resistance against insects and disease. Today, some farmers are growing organic corn without the use of herbicides and are cultivating the corn to control weeds. This practice is almost identical to the standard method of growing corn prior to the 1940s.

During the 19th century farmers grew Indian corn, a hard flint corn that produced an average of 30 to 40 bushels of ear corn to the acre. Hybrid corn of the dent variety came into use in the 20th century and shelled corn yields have gradually increased to 130 bushels per acre.

Late in the 19th century the silo appeared on the farm scene and gradually changed the way cattle

Corncrib, circa 1950.

were fed. The corn was cut by hand in the field, while still green, and was drawn by horses and wagon to the silo where the whole stalk was chopped and stored for cattle feed during the winter months. The first silos were wood, but in the 20th century masonry silos became common. With the advent of the farm tractor more and more farms had silos. A corn harvester was developed that cut the stalks at the base and tied them with string into bundles. The bundles were loaded on a wagon and drawn to the silo where an ensilage cutter, powered by a tractor, cut them into fine pieces and blew them into the silo. In the 1950's the field chopper came into use, chopping the corn in the field. The chopped corn was drawn by wagon to a blower and blown into the silo. This, coupled with an automatic silo unloader, eliminated all manual handling of the corn. Farms with a large number of cows now use bunker silos, which are less expensive to construct, easier to fill with ensilage and easier to remove the ensilage for feeding the cows. The bunker silo has replaced many of the vertical silos and eliminated the need for a blower and silo unloader.

The mechanization in farm corn production has reduced labor so one person can grow over 500 acres without hiring extra labor, but the equipment for its efficient production is so expensive that large acreages of corn are necessary. The strong market for corn encourages production, as do government programs supporting its price.

Corn acres have increased in New York and throughout the United States in recent years. It is being used to produce ethanol, an alternative to gasoline, as a fuel for motor vehicles. The extra demand for corn has raised its price, encouraging additional production. As other alternative sources for energy, such as ethanol from cellulose are improved, corn production is likely to decrease.

Grain production has been an important part of our New York agriculture for hundreds of years. Grain served as a major source of food for the Native Americans and continues to be an important food source today. The grain produced in New York today is largely fed to livestock with only a small portion used for direct human consumption. Most of the grain we eat comes from western states where it is processed into flour or the grain products we enjoy and shipped to our stores packaged, ready for consumption. Today man is consuming more grain than ever in the past, but the grain is consumed indirectly in our meat, eggs and dairy products. New York's climate and soils are conducive to the production of grain so we will, in all likelihood, continue to produce large quantities of grain long into the future.

Chapter IX
Livestock and Poultry

LIVESTOCK

Livestock, along with the products from livestock, are and have been the major New York agricultural enterprises. Going back to our first New York census of 1825, we find the number of cattle, horses, sheep and hogs listed for each of our counties. Oneida County ranked first with over 60,000 head of cattle followed by Otsego with 56,000. Montgomery County ranked first in horses, with over 15,000, but at that time it included both Fulton and Hamilton counties. Dutchess County ranked first in sheep, with over 174,000, and also was first in hogs with 78,000. **Table 17** provides the number of various livestock in New York at several different census dates.

The early census figures refer to cattle as "neat cattle". This is an all-inclusive term that includes all cattle and calves regardless of their intended use. It also includes oxen, which are not specifically mentioned until the U.S. census of 1850, showing over 178,000 in New York. At that time the state with the next greatest number of oxen was Missouri with 112,000. In the early 1800s cattle were all-purpose, meaning that the same breed was used for both milk and beef production. These all-purpose cattle were not large milk producers. The male was either castrated for use as an ox or fed for beef. The 1840 U.S. census was not the full agricultural census that was conducted in 1850 and each decade thereafter, but did ask a few agricultural questions including

TABLE 17 – New York Livestock Numbers
(in thousands) NY and US Census

Year	Cattle	Horses	Sheep	Hogs
1825	1,513	350	3,497	1,468
1845	2,072	505	6,444	1,584
1875	2,250	594	2,119	960
1900	2,596	628	1,746	677
1925	1,837	440	473	259
1950	2,022	136	155	168
1978	1,620	52*	66	126
1997	1,450	47*	61	79
2007	1,443	85*	63	86

* Horses & ponies

Three horse team on binder circa 1920. Note the fly nets, made out of rawhide strips, on the horses. As the horses moved, the strips moved, brushing off the flies. Courtesy of the La Fayette Historical Society.

A flock of sheep and lambs in a pasture at Bristol Center in Ontario County, circa 1913. Courtesy of New York State Archives.

Piles of fleece wool in storage at Geneseo in Livingston County, circa 1913. Courtesy of New York State Archives.

one about the number of cattle. In 1840, New York had almost two million cattle and Ohio was second with one-third less.

In 1840, New York had the greatest number of horses followed by Ohio, which in 1850 surpassed New York in the number of horses. The advent of the tractor on our farms brought about the rapid decrease in horse numbers. During the last half of the 19th century, as New York forests were cleared and more land became available for crops, the number of oxen gradually decreased, while the horse population increased.

Orange County had over 12,000 working oxen in 1850, almost double that of any other county. From the beginning of settlement, the cow was important, not only for food, but also for producing bull calves that were castrated and trained as oxen. It was oxen that pulled the trees of the forests into piles for burning, pulled the stumps out of the fields, provided the power for plowing the fields and pulled the sleds and wagons carrying the crops to market. Finally, when the ox could no longer labor, it provided meat for the table. The ox has vanished from New York's landscape but was instrumental in the State's development.

The years prior to 1850 were the age of homespun with clothing made, largely in the home, from wool and flax. The 1845 census shows over six million sheep in New York, more than two sheep per resident. Sheep numbers dropped dramatically after 1850, as more clothing was produced in factories and there was a greater use of cotton. During the Civil War, the number of sheep increased for a few years when cotton was not readily available from the south.

There has been a small increase in the number of sheep during the last decade. There are quite a few people spinning wool as a hobby or as a small business. Tom Clark, of Old Chatham, has several hundred sheep that are milked in a parlor especially designed for sheep, and uses the milk to make cheese.

Pork was a staple food in the homes of most New Yorkers in the early 1800s, as hogs obtained much of their food from foraging in the forests. When states to the west were settled after the opening of the

Hog butchering, 1920. Several farmers often got together to butcher their hogs for their year's pork supply. Courtesy of La Fayette Historical Society.

Erie Canal, and grain became the major hog food, the numbers of hogs in New York decreased with pork production moving to the Midwest. Hogs were often grown adjacent to breweries and distilleries and fed from their by-products. It was a common practice for many farmers, until the mid-1900s, to grow two or three hogs each year and butcher them in the winter for the farm family's use. With the trend toward specialization, few farm families grow their own pork today.

The majority of the cattle in New York have been raised for the production of milk and its by-products. Numbers of beef cows are not listed until the 1917 census, which shows a little over 39,000 beef cows in the State compared to over 1,300,000 dairy cows. Beef cow numbers decreased to 12,540 in 1940 but since then have gradually increased to 103,620 in the 2007 census. Many farms sold their dairy cattle between 1940 and 2007. Usually a farmer selling his dairy had a barn, pastureland and equipment for producing hay, so he could easily keep some beef cattle without having to update his equipment and not be tied to the twice a day milking regimen.

Although all cattle are grouped together until into the census figures of the 20th century, we know that beef was an important part of the earlier settlers' diet by data from the U.S. 1810 Census of American Manufacturers (**Table B in the Appendix**). This census lists 867 tanneries in New York with 151,165 hides and 210,445 calfskins tanned that year. Animal hides were an important by-product of beef production. Tanneries purchased the hides to make leather for shoes and other leather needs of the early settlers. With 867 tanneries scattered around the State, it was convenient for the farmer to have the hide of the animal, which was slaughtered on the farm, processed locally.

In the early days of the State, an itinerant shoemaker went from farm to farm, often using the leather from one of the farms animals to make shoes for the family. As the population increased, a shoemaker would find full-time work in a village and make shoes to order. You will note in **Table B** of the Appendix, there were 867 tanneries in the State in 1810. Like almost all industries the tanneries grew in size, moving to larger population centers where shoes were mass-produced.

DAIRYING

The 1855, New York Census lists the number of cattle killed for beef as 225,338. That same year there were a total of 1,213,024 cattle noted in the census. Since the cows were multi-purpose for both

TABLE 18 – Dairy Cow Numbers, Butter and Cheese Production on the Farm and Whole Milk Sales
NY and US Census Figures

Year	Dairy Cows (000s)	Butter Produced (000s) Pounds	Cheese Produced (000s) Pounds	Whole Milk Sold (000s) Pounds
1845	999	79,502	36,745	NA
1855	1,068	90,293	38,944	180,306
1875	1,340	107,873	7,778	357,000
1900	1,502	74,714	2,625	3,830,680
1925	1,370	22,106	NA	4,754,115
1950	1,218	NA	NA	7,480,811

NA - Not available

dairy and beef, it is likely thta close to half of those slaughtered were retired milking cows.

New York was by far the leading dairy state at the time of the first U.S. agricultural census in 1840. The value of New York's dairy products that year was $10,496,021 compared to Pennsylvania, which was second, with less than a third of New York's value. In 1850, New York produced over one-fourth of the butter and almost one-half of the cheese produced in the U.S. By 1900, Iowa surpassed New York in butter production but New York was at the top in cheese production producing over 43% of the nation's total production. By 1950, New York had moved into second place behind Wisconsin in the value of its dairy products. Today, California is in first place followed by Wisconsin and New York. Dairying, today, generates over half of all New York State agricultural income.

The 1825 census shows over a million and a half cattle in New York and by 1845 the number surpassed two million, with just under one million of those dairy cows. **Table 18** shows the number of dairy cows, the amount of butter produced on farms, the amount of cheese produced on farms and the quantity of whole milk sold.

In 1845, Orange County was the leading producer of butter with 4,109,000 pounds followed by Oneida County with 3,876,000 pounds. Herkimer was the leading cheese-producing county with 8,209,000

TABLE 19 – Number of New York Farms, Farms with Dairy Cows, Dairy Cows and Average Pounds Production of Milk per Cow
US Census

Year	Number of NY Farms	Farms with Dairy Cows	Total Number of Cows	Average Milk per Cow
1900	226,720	196,366	1,501,608	4,426
1925	188,754	151,371	1,370,060	4,734
1950	124,977	87,736	1,217,596	6,810
1969	51,909	24,775	898,147	10,682
1978	43,075	18,493	850,381	11,488
1997	31,757	8,732	700,480	16,519
2007	36,352	5,683	626,455	19,303*

* New York Agricultural Statistics

pounds followed by Oneida with 3,278,000 pounds. It is logical that Orange County, relatively close to a large metropolitan area, ranked first in butter because the lack of refrigeration prohibited shipping butter a long distance.

By dividing the pounds of butter produced by the number of cows, we find that New York farmers were making an average of 80 pounds of butter and 37 pounds of cheese from each dairy cow during the years 1845 and 1855. By 1875, the amount decreased to an average of about 70 pounds of butter and 6 pounds of cheese per cow. In the 1860s, the production of cheese began to move to cheese factories located in many rural communities. The 1875 census shows over 83 million pounds of cheese made in factories, the equivalent to an average of 62 pounds of cheese per cow. Creameries for the production of butter were beginning to operate in 1875 and produced a little over 3 million pounds that year, only about 3% of the amount produced on farms.

Whole milk sales, prior to 1875, were limited to farms close to villages and cities. There was no refrigeration other than block ice cut from lakes and ponds and few families were fortunate to own an icebox. Whole milk sales increased substantially between 1875 and 1900 but much of it was going to creameries for the production of butter as well as for the sale of whole milk to the consumer. Milk was not the universal beverage in the 19th century that it became in the 20th century.

Table 19 shows the changes in the number of New York farms, the number of farms with dairy cows, the total number of cows and the average milk production per cow.

There are about 16% as many farms today as there were in 1900 and about 3% as many farms with dairy

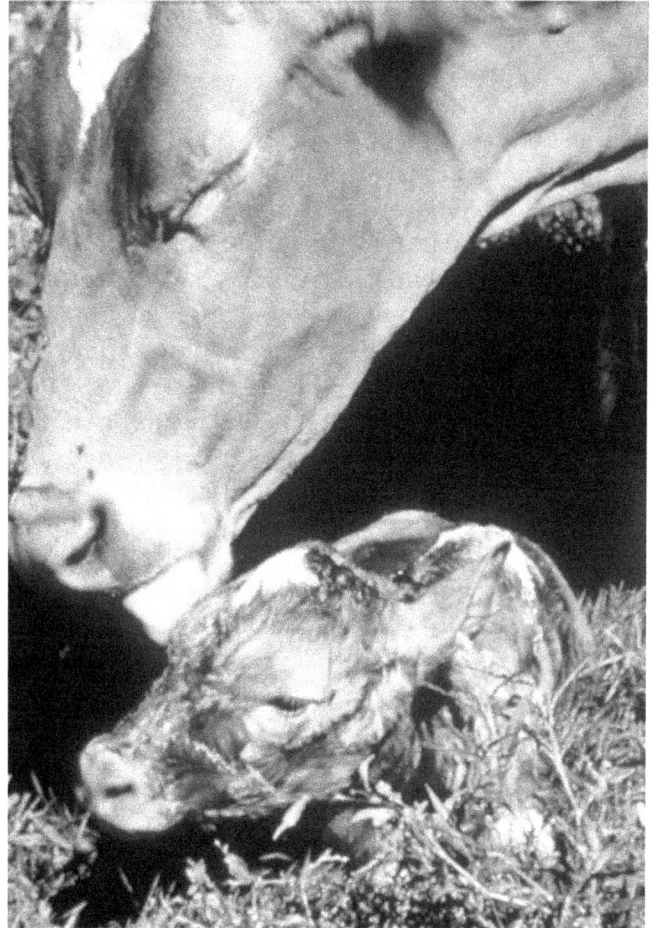
A Guernsey cow with newborn calf, circa timeless!

cows. The dairy farms remaining are producing over twice the milk with less than half the cows of 1900. These are dramatic and significant changes.

Both the decrease in the number of farms and the decrease in the number of farms with dairy cattle have been catastrophic to thousands of New York farm families. A livelihood that existed for generations disappeared, with many farm families forced to discontinue farming and find alternative occupations. In only 44 years, between 1925 and 1969, New York lost 72% of its farms and 83% of the farms with dairy cows!

Up until the early 1900s, the large majority of New York farms had at least one dairy cow to provide milk for the use of the farm family. From the middle of the 1800s, with a growing population and railroads to furnish rapid transportation, dairying became a good source of income for New York farmers. **Table 20** shows the percentage of New York farms

TABLE 20 – Percent of New York Farms with Dairy Cows and Average Number per Farm

Year	Percent of Farms with Dairy Cows	Average Number Cows per Farm
1900	86.2	7.6
1950	70.2	13.9
2007	15.6	110.2

79

with dairy cows and the average number of cows per farm.

Mechanization was the prime force making specialization necessary because of the additional cost of owning the necessary equipment to make a farm operation efficient. As mechanical innovations and improvements came to the farm, each generation, since the time of the Civil War, has had to make the decision whether to make the necessary changes or discontinue farming. The introduction to the 1875 New York census stated, "there is a marked tendency toward the absorption of smaller farms by larger farms". When the local cheese factory opened, the farmer had to decide whether to add a few more cows in order to make the change profitable, or continue making cheese on the farm. When the local creamery opened, he had to decide if it was appropriate to construct an icehouse and cut ice to cool his evening milk. As the years progressed there were more questions he had to answer. Shall I enlarge the barn? Shall I construct a milk house? Should I build a silo? Can I afford to buy a baler? Should I install a bulk tank? Can I afford to build a milking parlor and a free stall barn? Shall I put in a bunk silo to replace my vertical silos? These are a few of the questions dairy farmers have had to ask themselves over the last several generations. The ones whose answers were "no" were part of the thousands that decided that someone else should be in dairy farming.

There were many other reasons for farms to discontinue dairying. They included the size of a farm and the lack of opportunity to increase its size, poor quality of the farm's soils, lack of capital, preference for other work, age, no interest in farming by the children and land becoming surrounded by development. These reasons have always been logical ones, but the main force contributing to the decrease in the number of dairy farms was the necessity to mechanize and expand to remain a viable operation.

Dairying has been such an important part of New York's agriculture from its beginning, 200 years ago, to today that a description of some of the changes that have occurred is appropriate. In the mid-1800s the majority of farms produced either butter or cheese or both. Butter and cheese were consumed

by farm families, with the surplus sold to stores or dealers in the local community. Each farm that sold butter had a distinctive wooden stamp or wooden mold with a special design to mark the butter so the consumer could tell who made it. A reputation for excellent butter would bring a better price. It was an early form of brand marketing. The butter markers had attractive designs and are very common today in museums and at antique shows.

Initially, the cows foraged in the forests eating the plants and grasses that nature provided. Both the quantity and the quality of the milk left much to be desired. At milking time the cow was tied to a tree and the housewife milked the cow in the open air. Later a barn was constructed and the cows were milked in stanchions. The cows were bred in the summer to freshen in the spring when the new grass supplied them with food. This also gave a dry period in the winter when the cows did not have to be milked. It also gave the calves a better opportunity to reach an age where they could forage the following fall and winter.

Milk did not become a major beverage until the 20th century. Until electricity arrived it was difficult to keep milk cold enough to prevent spoiling. Most of the milk was turned into butter and cheese, both of which had a much longer shelf life.

Cheese had a longer shelf life than butter and could be stored and transported longer distances without refrigeration. Cheese factories came into existence soon after the railroads came to many parts of New York. Cheese had been shipped on the canals, but it was a lengthy process if the markets were some distance away. Buyers found a great variation in

TABLE 21 – Butter and Cheese Reported as Produced on the Farms and Factories of New York
1900 US Census

Butter reported by farms	74,714,376 pounds
Butter reported by factories	40,693,846 pounds
Total reported	115,408,222 pounds
Cheese reported by farms	2,624,552 pounds
Cheese reported by factories	127,386,032 pounds
Total reported	130,010,584 pounds

TABLE 22 – Gallons of Milk and Cream Reported Sold by Farmers and Milk and Cream Reported Purchased by Factories of New York

1900 US Census

Sold by Farmers		Purchased by Factories	
As Milk	445,427,888	As Milk	271,717,600
As Cream*	3,354,263	As Cream*	1,674,636
Total	448,782,151	Total	273,392,236

* Expressed in term of milk, 1 gallon of cream to 5 ½ of milk

the taste, smell and quality of the cheese from one farm to another. To overcome this problem cheese factories were constructed in almost every farming community during the last half of the 1800s, buying milk from many farmers and turning out a large quantity of a uniform cheese. The railroads provided a rapid means of moving the cheese to markets.

Table 21, with information taken from the 1900 U.S. Census provides an informative picture of the changes taking place in New York's dairy industry.

One can observe, by looking at the above table, that butter production in the factories was a relatively new and growing industry, whereas the production of cheese in the factories was a mature industry having started approximately 50 years earlier.

Creameries for the production of butter came some years after the first cheese factories. There were numerous creameries throughout the State during the late 19th century and early in the 20th century. Butter making was gradually moving from the farm to the creamery, a trend that continued until the

The Windhausen Family, ready to milk the cows, circa 1912. It was common for farm children to help with the milking. Courtesy of Jamesville Community Museum.

1940's. Instead of each farm having three or four cows, some sold their cows while others added a few more when they made the decision to sell their milk to a creamery.

Tables 18 and 22 give us a good picture of how the milk, sold by the farmers in 1900, was being used. In addition, since it takes approximately three gallons of milk to make a pound of butter and approximately 1.2 gallons to make a pound of cheese, the two tables show that by 1900 more milk was being sold from the farm as whole milk than was being used on the farm to produce butter and cheese. Any milk and cream not sold to the factories was sold directly to the consumer while some was sold to milk dealers for resale to the consumer.

The majority of farmers milked their cows by hand until the 1930s even though mechanical milking machines had been invented a half-century earlier. The arrival of electricity on the farm to power vacuum pumps speeded the adoption of the milking machine. Hand milking was a labor-intensive job that had to be done twice a day, every day. This kept the size of the herds small until there was mechanization. When there were two or three cows it was common for the housewife to do the milking and make the butter. As the number increased it often became the responsibility of the farmer to do the milking.

When butter was made in the home, milk was put in pans in the cellar, where it was cooler, for the cream to rise. Butter would be made whenever enough cream had accumulated and before it had soured. The skim milk, which would have a limited amount of fat, could be made into cheese or was more often used to feed hogs. Some farms had a spring that flowed continuously and constructed a special building, called a springhouse, directing the flowing water into shallow tanks. They set pans of milk in this cold running water for the cream to rise. Few farms were fortunate enough to have such a good supply of continuously running cold water.

Only a few farms in the State had electricity before the 1920s and many did not receive electricity until about 1940. When a farmer decided to send his milk to a creamery he had to cool the evening milk to keep it from spoiling, but could take the morning milk without cooling, if it was delivered to the creamery in a timely manner. The only method

of cooling the milk, for most farmers, was to set the milk cans in tanks of ice water. Ice was cut from lakes or ponds in the winter and stored for use during warmer weather.

Icehouses were the solution for this need. Ice was cut on lakes, rivers and any accessible pond. The farmer constructed a building close to his dairy barn and filled the icehouse studding with sawdust for insulation. The ice was cut into blocks with special saws, loaded on sleighs and drawn by horses to the icehouse. More than one farmer had a cold dunking in the water while harvesting ice. The ice was packed tightly together with sawdust put around the outside and on top. When warm weather arrived a cake of ice was removed each evening and put in a water vat along with the cans filled with milk.

Construction of an icehouse was an extra expense and coupled with the fact that harvesting ice was hard, difficult work, a farmer might decide not to add cows to produce milk to send to a creamery. Instead, some farmers continued to produce butter in the home rather than make the change, but gradually these farmers discontinued milk production.

When the farmer decided to discontinue making butter on the farm and send his milk to the creamery he needed a place to store his milk overnight. The cellar of the home was no longer convenient so a milk storage building, called a milk house, was constructed. Sometimes a farmer would attach a small addition to his barn but usually built a small building nearby. A vat was constructed of concrete to hold the milk cans along with the water and ice to cool them. If there was a readily available supply of water, a wash vat might be put in the milk house to wash the milking utensils. If hot water wasn't readily available, the farmer continued to take the milk utensils to the house for daily washing.

In the early 1900s, housewives could take their milk container out to the street as the dairyman came by with his horse and wagon, to purchase their milk as it was dipped out of the farmer's milk can. The icebox in the kitchen extended the life of the fresh milk. When pasteurization arrived in the early 1900s, fresh milk could be kept in the icebox even longer. Although pasteurization of milk had been suggested in 1886 it did not become common until dairy processing moved to communities where electricity was available and the size of the

Calf hutches on a Cayuga County dairy farm in 2009.

dairies permitted purchase of the relatively expensive pasteurization equipment. In 1931, the State mandated that milk sold by milk peddlers had to be pasteurized.

By 1920, with the development of paved roads and larger trucks, milk from many farms was being transported to larger milk processing facilities in villages and cities and to large plants located on a railroad. A custom hauler contracted with the large dairy processor to pick up the milk from a number of farmers who shipped milk to that plant. The milk was transported in 10-gallon cans with an assigned number painted on the sides and top of each producer's cans.

Almost every village in the State had one or as many as three small dairies that peddled milk to the consumers in the village. Often a farmer decided to market the milk from his cows in this manner to increase his return. If demand exceeded the production from his cows he purchased milk from other farmers. Some farm dairies became quite large purchasing milk from a number of other farmers and selling it in several villages.

An important implementation of new technology occurred in about 1938 with the formation of the New York State Artificial Breeders Cooperative (NYABC). Artifical breeding of dairy cattle was in its infancy at the time and with support from the Cornell Dairy faculty the Cooperative was formed and a facility for the bulls was constructed not far from the Cornell campus. The Cooperative purchased excellent bulls that had been proven to produce superior progeny, artificial insemination technicians were trained and semen from these bulls was made available for breeding the cows of the Cooperative members. It was common practice, prior to the establishment of the Cooperative, for the farmer to save one of his own bull calves to utilize for natural breeding and very often the bull did not contribute the best characteristics needed for a high-production dairy cow. The utilization of artificial breeding helped raise the average production per cow on New York farms.

With the aid of Cooperative Extension, a number of counties formed Dairy Herd Improvement Associations (DHIA) prior to World War II. In 1949, they gradually joined together to form a New York DHIA.

It was designed to help farmers improve their dairy herds and to improve their profitability. A representative of the DHIA came to the farm once a month to weigh the milk and test the amount of butterfat of each cow. A farmer could compare the production of each of his cows to determine from which cows heifer calves should be saved, and also to compare the test results of his herd with other herds in the county. The farmer paid the DHIA a fee for each cow that was tested. The New York DHIA was reorganized in 1966 and later became associated with DHIA of other states becoming the Northeast Dairy Herd Improvement Association.

In the 1950s, bulk milk coolers came into use on farms with the custom milk hauler pumping the milk directly from the farmer's refrigerated tank to the hauler's refrigerated tanker truck. Initially, many milk processors offered a premium to a farmer to encourage installation of a bulk tank. The additional cost of the bulk tank, along with the cost of a building to house the tank, helped many dairy farmers make the decision to discontinue dairy production.

Milk production per cow increased dramatically over the years because of improved animal nutrition and breeding for increased production. The cows are also larger than they were in the past. In the 19th and early 20th centuries the amount of butterfat in the milk was very important. In recent years consumers are choosing low fat milk making both protein and quantity more important.

In the 1990s, a number of farmers gave their milking cows additional amounts of a product called BST, a naturally occurring hormone in the cow, which increases milk production. There has been some resistance by consumers to the purchase of milk from cows that receive this additional hormone.

Devons and Durhams, also called Milking Shorthorns, both multi-purpose cows for beef and milk, were the first breeds. As milk production became more important new breeds were developed or imported. In 1885, Smith and Powell Dairy Farm near Syracuse imported Holsteins, an excellent milk producing breed, from Holland. They brought in 1,283 head of cattle, which was one eighth of all that were imported into the US.

Onondaga County was famous as a source of fine Holstein cattle in the early 1900s. In addition to Smith and Powell, there was the Brookside herd of the Stevens Brothers, Moyerdale owned by Harvey Moyer in Salina, Maplehurst in Manlius and other large dairy herds in the State. There was a large sales facility in Liverpool, in Onondaga County, with buyers coming from many states and even foreign countries to purchase fine cattle.

In recent years, many consumers have chosen to purchase natural and organic farm products produced without the use of manmade chemicals, including inorganic fertilizer and pesticides. A number of dairy farmers are supplying consumers with organic milk. The demand has been steadily increasing during the last decade providing an excellent market for these dairy farmers.

The 2007 U.S. census shows there were 142 dairy farms in New York milking over 1,000 cows and another 145 milking between 500 and 1,000 cows. Wyoming County has the largest number of milking cows with 47,970, followed by Cayuga, St. Lawrence and Jefferson all with a little over 30,000 cows. Most of these farms milk their cows three times a day with around-the-clock milking. The cows are stabled in large, free-stall barns with a complete mixed ration of hay, silage and concentrate available much of the time. At milking time they are moved to the milking parlor, where the cows' udders are cleaned and milking units are attached, which are automatically removed upon completion of milking. The milk passes through sanitary piping into a quick cooling system and then is stored in a refrigerated bulk tank until the daily pick up by a refrigerated bulk truck for transport to the milk plant. On some of the large dairy farms, the milk travels directly into the refrigerated tank of a tractor-trailer, ready to go to the milk plant. In recent years, a large portion of the labor on the bigger farms is from Hispanic neighboring countries to our south. The unsettled immigration policies of our country may make this labor supply more difficult to obtain in the future.

There are continual innovations to increase the labor efficiency of dairy farming. One of the most intriguing ones involves equipment that automatically milks the cow without any person present. When the cow feels it is time to be milked, she goes to a special area where she receives some feed, her udder is washed, the milking machine is attached, milking occurs, the milking machine removed and the gate is opened for the cow to leave after being

A rotary milker for 72 cows at a time on Merrell Farms, Wayne County, 2009.

milked. All of this is done robotically. There has been at least one of these units successfully operating near Rochester for several years, and the Maple Lawn Dairy, owned by the Wolf family of Lyons, has had five units operating successfully for a year. Each robotic unit services 50-55 cows and several other farms are in the process of installing robots.

The milk from smaller herds is usually piped directly from the milking machine to the bulk tank, but the cows are often milked in their stalls, usually twice a day. A number of these smaller herds graze on rotational pastures during the late spring, summer and early fall months. Although the recent trend has been toward larger sized herds, small, well-managed herds are still operating profitably. The farms with 100 or fewer milking cows have an advantage in that most of the labor is furnished by the farm family, but have a disadvantage in that the family has to milk twice-a-day, 365 days a year.

Our New York dairy industry should remain strong well into the future. We have the right soils and climate to produce feed for cows that is both economical and nutritious. Recently, because of the cost of energy and consumer desire to buy food close to where it is produced, demand has increased for locally grown food. This trend will likely continue and with 19 million people in our state and millions of others only a few hours away, demand for milk from our farmers should increase.

Goats have had a presence on farms for many years, and in the 19th and early 20th century they were often trained to pull a cart for a child. They have been a minor source of both milk and meat, but recently more are being raised. The 1940 U.S. Census lists 2,859 farms with a total of 9,425 goats. The 2007 U.S. Census shows 2,707 farms with 39,920 goats. Ethnic populations, that commonly enjoyed goat meat in their native lands, have created a demand for goat meat here that farmers are meeting.

Feeding the chickens by scattering grain on the ground, circa 1900. Courtesy of the Marcellus Historical Society

In 1950, there were 5,352 farms with rabbits; in 2007 that number dropped to 1,100 farms. Rabbits are becoming more popular as a source of meat, rather than for their fur, as was the case in the past. There were 414 farms producing furs in 1950 and there are only 5 farms listed as producing furs in 2007. There were several very large mink and other fur producers in the 1950s and 1960s. Excess production in Russia and other areas of the world lowered prices to an unprofitable level forcing most of the fur producers to go out of business. Protests by animal rights groups against using animal furs for clothing also decreased the market for furs.

There are several other varieties of animals produced by New York farmers, as listed in the 2007 Census. There are 67 farms with 1,854 bison, 180 farms with 7,847 deer, 42 farms with 1,351 elk, 506 farms with 7,100 alpaca and 465 farms with 2,393 llamas. Many of these farms are part-time farms where growing these unusual animals is a hobby.

It is difficult to consider bees as livestock but they are placed in the livestock category in the U.S. Census figures. Honey had been a popular sweetener for thousands of years and was enjoyed by our early New Yorkers as well as our residents today. The 1855 New York Census lists the honey production for that year as a little over 2½ million pounds, along with 138,000 pounds of beeswax. Production has varied from year-to-year but in 2007 we produced over 3 million pounds of honey from 805 farms with 46,401 hives of bees. Both the number of hives and the production of honey have decreased by about 25% between 1997 and 2007. New York has a large fruit industry and bees play an important part in fruit production by fertilizing the fruit blossoms. Fruit farmers hire the beekeepers to bring hives of bees to their fruit orchards each spring to fertilize the fruit blossoms. There has been an unknown agent, perhaps a virus, infecting and destroying many hives of bees in recent years. Research is in progress to find and eliminate the cause.

POULTRY

During the 19th century and into the first quarter of the 20th century, almost every New York farm had poultry. The poultry was largely chickens with the pullets saved for egg production and the cockerels designated for an early visit to the cooking pot. Many farms also had a few ducks, geese, turkeys or guinea hens. These birds usually ran loose around the farmyard enjoying grass and insects. The 1900

Census illustrates the extent of poultry, indicating that of 226,720 New York farms, 206,389 had poultry. The trend toward specialization in poultry on the farm becomes quite evident, in the 1950 Census, with poultry on only a little over a third of New York farms.

The first poultry records come from the 1840 Census. It shows New York leading the states in the value of poultry at $1,153,413, with Virginia in second place with only two-thirds as much. This was not surprising as at that time New York had more farmers than any of the other states. The 1855 New York Census lists the value of poultry sold that year at a little over a million dollars, and the value of eggs sold at one and a third million dollars. In 1900, the value of poultry raised was over $6 million and there were 62 million dozen eggs produced.

Very often the chickens were the farm wife's enterprise. She took care of them and the money received from the sale of eggs was hers to use. The eggs were often traded, at the nearest general store, for household items.

Nature designed birds to lay their eggs in the spring. By collecting eggs from the nest daily, the chicken continued to lay eggs into the summer. The farm wife discontinued collecting the eggs in the summer, before the chickens stopped laying. The hen continued to lay until there were about 15 eggs in the nest before she started setting on them for hatching. Once she started setting on her eggs, she discontinued laying until the next spring. To have eggs in the fall and winter, the farm wife put eggs into large clay crocks containing a preservative called water glass. She removed eggs from the crock whenever she needed them for baking.

The incubator came into use in the 1880s and as its use gradually spread, it eliminated the need for the hen to set on her eggs for hatching. Some farms installed an incubator and sold the day-old chicks to other farmers who grew the chicks, all at one time, in brooder houses. Later, specialized personnel at the hatchery were able to separate the female chicks from the cockerels eliminating the need for egg-producing farms to grow chickens for meat.

With the advent of electricity on the farm, lights in the chicken house were turned on at night, fooling the chickens into thinking it was spring. By keeping the hours of light the same as in the spring, chickens lay eggs all year. Occasionally a poultry farmer purchased a small gas generator to light the chicken house before electricity came to the farm and before it was available in the home.

A large chicken laying house, circa 1950.

TABLE 23 – Value of Poultry Sold, Number of Broilers & Ducks Sold and Layers on Farms
from US Census

Year	Poultry Sold in $000s	Broilers Sold (000s)	Ducks Sold (000s)	Layers on Farms (000s)
1940	32,199	NA	5,368	11,190*
1950	80,425	NA	4,217	10,352*
1997	106,620	1,320	NA	3,802
2007	123,727	1,780	2,432	3,953

*Chickens over 4 months of age

Chickens were provided a building with a small door on one side that could be opened in the morning permitting the birds to run loose, eating grass and bugs, which provided much of their diet. They would be shut in the houses at night to protect them from owls, foxes and stray dogs. As the industry began to specialize, in the mid-20th century, chicken houses became larger and the laying hens were confined to the building. With improvements in housing, breeding and nutrition, today a laying hen produces approximately 300 eggs a year.

The political slogan of "a chicken in every pot" in the early 1900s signified the value of chicken meat compared to the red meats. Chicken meat was more expensive to purchase because, with eggs produced only in the spring, the cost of production was greater. When a family had special guests, chicken was served. With the advent of artificial lighting in the chicken house, chicken production increased and the cost of the chicken meat decreased. Scientists developed nutritionally balanced feeds, which made the chickens grow faster with less feed and also made it possible for the chickens to be grown in large buildings without the previously necessary pasture, bugs and sunlight. With these advancements, poultry breeders developed chickens designed either to lay a large number of eggs or to produce meat efficiently. Prior to the middle 1900s, chickens were bred to be multi-purpose, for both egg and meat production. In the middle 1900s, breeders bred smaller chickens that would produce large eggs with less feed and also larger chickens that grew rapidly and were used in meat production. These specialized breeds helped decrease both the cost of eggs and the cost of chicken meat to the consumer.

Table 23 shows the value of poultry sold, the number of broilers and ducks sold and layers on farms for several periods of time.

Hudson Egg Farm in Onondaga County, 2008.

Turkeys on a pasture at Plainville in 1939. Courtesy of the Liverpool Library's Schuelke Collection.

Poultry has become a larger portion of meat consumed since the advancements in poultry nutrition and breeding. Consumption of beef and pork per capita has changed very little since 1960, while the consumption of chicken and turkey has tripled. Most of the chicken broiler, duck and turkey production has moved from New York to other states where there are lower costs. There are still a large number of broilers grown in Sullivan County for the kosher market and recently production of kosher chickens and turkeys started in Onondaga County. Long Island was noted for many years for its duck production, but as urbanization pressures have increased, the production of ducks on Long Island has decreased.

In **Table 23** it might appear that chicken egg production has decreased more than it actually has. The figures for layers in 1940 and 1950 were not available, so I included figures for chickens over four months of age. These figures would have included some chickens not yet into production, capons and roosters. Another factor to consider is that production of eggs per hen has increased substantially since 1940.

In 2007, just 20 of the 3,345 farms with layers produced over 95% of New York's eggs and 8 farms produced 85%. The number of farms with poultry is high because there are many farms that grow unusual varieties of chickens as a hobby and, in addition, there has been a large increase in the number of farms growing a few chickens for their own family's consumption. A third category, that is increasing, is the number of farms growing poultry for organic meat or organic egg consumption. A large portion of all these farms has less than 50 layers.

The large egg-producing farms have state-of-the-art operations. The production of poultry and eggs has become very competitive, requiring businesses to

Duck farms at Speonk on Long Island, circa 1919. Courtesy of New York State Archives.

become as efficient as possible. The laying chickens are housed in large buildings with the light and temperature closely controlled. Eggs automatically roll onto a belt and are transported to the packing room where they are washed, candled and packed. Some people feel that because the chickens are caged they are mistreated, but that is not the case. For a farmer to obtain the greatest production from his poultry or livestock, it is necessary to make them as happy as possible. I have been in large chicken houses and it is obvious to anyone who knows poultry, that the chickens are happy and healthy.

In 1949, there were 671,240 turkeys grown on 1,128 New York farms with Jefferson County growing the most with 74,208. By 1954, the number of farms with turkeys had increased to 2,742 with over 900,000 raised. Just five years later, only one-third the number of farms grew 800,000 turkeys. This is a good example of the consolidation of agriculture during the 1950s. By 2007, the number of farms with turkeys had decreased to 313 and over 90%

of turkeys were being grown at Plainville Farms in Onondaga County.

There were a number of poultry breeders scattered throughout New York during the first half of the 20th century, but like other segments of agriculture, the poultry breeders gradually went out of business or consolidated into large, more efficient enterprises. Breeders of poultry located closer to the large production areas of the US where cost of production was less.

Today, commercial poultry production in New York is limited to a relatively few farms that are producing for specialized markets, rather than the small flocks of poultry on almost every farm a century ago that supplied the majority of poultry for New York residents. There will continue to be niche markets for organic, free range and kosher poultry, but the majority of the poultry products consumed will come from outside the State.

Chapter X
Vegetables

Few New Yorkers recognize the importance of New York-grown vegetables to both farmers and consumers. In the annual bulletin, *2007 New York Agricultural Statistics*, published by the Department of Agriculture and Markets, New York's rank in production among the 50 states for fresh market vegetables is: second in cabbage; third in cauliflower; fourth in pumpkins, snap beans and sweet corn; fifth in squash and sixth in cucumbers. The fresh vegetable production of New York encompasses over 82,000 acres and another 78,000 acres of vegetables are grown for processing. The U.S. 2007 Census ranked Genesee County first, with 27,220 acres of vegetables and Orleans second with 18,914. Orange County leads in the production of onions.

The 1840 U.S. Census ranked New York first in the US with a value of $499,126 in market produce followed by Massachusetts with $283,904. By 1850, the value of New York's market produce had increased to $912,047 with Queens County producing $308,957 of that total. The majority of vegetables, at that time, were being grown in close proximity to the population.

Today's acreage of vegetables is in marked contrast to the 12,591 acres of New York vegetables listed in the 1855 census. At that time, transportation of fresh vegetables was difficult and there was no refrigeration to maintain freshness. As transportation improved and refrigeration became available, both the acreage of fresh vegetables and the per capita consumption increased. By 1900, New York grew over 100,000 acres of vegetables and in 1950 production surpassed 200,000 acres as shown in **Table 24**. Since 1950, acreage has decreased

TABLE 24 – Number of Farms Growing Vegetables, Acres and Value from Census

Year	Farms Growing Vegetables	Acres	Value ($000s)
1855	NA	12,591	499
1900	NA	106,738	7,850
1920	35,857	131,648	21,975
1950	15,919	206,830	37,154
1978	3,499	154,666	105,454
1997	2,720	169,331	206,866
2007	3,192	160,596	338,000*

*Includes potatoes

Workers picking a crop for a canning factory at Fayetteville, circa 1910. Courtesy of the OHA Museum & Research Center.

A transplanter with three men and a team of horses planting cabbage or tobacco. The barrel held water with about a cup dropping on the roots of each plant, circa 1910. Courtesy of the Museum at Shacksboro Schoolhouse, Baldwinsville, NY.

because of rapid transportation and the importation of fresh vegetables from all over the world.

A wide variety of the vegetables produced in New York are reflected in the 2007 Census. New York farmers planted 40,183 acres of sweet corn, 31,204 acres of snap beans, 13,618 acres of cabbage and over 18,000 acres each for potatoes and peas. Specialization during the last 50 years, where climatic conditions are ideal, has brought about a significant decrease in New York acreage of tomatoes, spinach, lettuce, celery, cauliflower, broccoli, cantaloupe and asparagus. The acreage of New York pumpkins has increased from 307 in 1950 to over 6,000 at the current time, while the acreage of sweet corn, peas, snap beans and cabbage has remained relatively stable. Pumpkins have been used in increasing amounts for fall decorative purposes.

The number of farms growing vegetables in 2007 was less than a tenth of the number growing vegetables in 1920 while, during this period, the acres of vegetables grown have increased. The economies of scale have made the cost of production for the small producer higher than the market is willing to pay, except for special situations where the produce fills a niche difficult for a large producer to fill. Two examples of this are the organic producer and the

TABLE 25 – Cases and Prices of Tomatoes and Corn Manufactured in New York
1900 US Census

Year	Tomato Cases	Average Price per case	Corn Cases	Average Price per case
1889	142,250	$.89	272,925	$.85
1899	254,616	$.81	1,341,352	$.70

Note: One dozen three pound cans of tomatoes in each case and one dozen two pound cans of sweet corn in each case.

small producer that sells retail in a green market or delivers specialty items to gourmet restaurants.

The canning industry had a feeble beginning in 1818 when Thomas W. Kensett, an immigrant to the US in 1815, started preserving foods in his New York City home. His father-in-law, Ezra Daggett, joined Kensett and in 1825, the firm of Kensett and Daggett was granted a patent for "an improvement in the art of preserving". Canning remained a very small industry for over a quarter of a century, partially because the cans were soldered by hand. In 1851-52, the Edgett brothers formed the first Upstate New York cannery at Camden. In 1866, one of the brothers built a canning factory at Newark, called the Wayne County Preserving Co. After this

date a number of canneries began to appear around the State, including Merrell and Soule of Syracuse who opened a cannery in 1869 specializing in the canning of milk products and mincemeat. [1] Merrell and Soule became a national leader in mincemeat production.

The commercial canning of fruits and vegetables moved forward more rapidly after the Civil War. Information from the 1900 U.S. Census, derived from the Census of U.S. Manufacturers, regarding the number

---

1   The Associated New York State Food Processors, *A History of the Canning and Freezing Industry in New York State* (1985) p. 17-19

Sweet corn is being husked in the tent and elevated by the conveyer into the canning factory, circa 1900. Courtesy of OHA Museum & Research Center.

of cases of tomatoes and corn manufactured in 1889 and 1899, is shown in **Table 25**.

The prices listed per case, in this census, went back to 1870 when the canning industry was nicely starting. In 1870, the canner received an average price of $2.10 for a case of tomatoes and $3.00 for a case of corn. Thirty years later the price of a case of tomatoes had dropped by over 60% and the price of a case of corn had dropped by over 75%. In 1899, the canner was receiving a little less than

seven cents for a three pound can of tomatoes and a little under six cents for a two pound can of corn. It is not difficult to see that the farmer received very little for his corn and tomatoes. The 1900 census, for the year 1899, stated that New York only had half a sweet corn crop that year, but managed to pack over a million cases.

The predecessor of our current tin can was called the "sanitary can" and came into use around 1900. It was mass produced and had a top the same size

Snap bean picker, circa 1963. Photo courtesy of New York State Archives from New York State Ag & Mkts.

Two snap bean harvesters in Onondaga County, circa 2009. Each harvester is picking four rows at a time. It would have taken well over 100 people picking beans by hand, as was done in 1950, to pick as many beans as one of these machines.

Loading cabbage for use in sauerkraut, circa 1970. Photo courtesy of New York Archives from New York State Ag & Mkts.

as the bottom making it possible to fill the can with items like sliced pineapple. Previously, the hand-soldered cans had tops much smaller than the bottoms. Until the time that cans were made by machine, the handmade cans were made by cannery employees and stored until needed for the canning season. It is not surprising that leaky cans were a problem for the early canning factories.

The New York Canned Goods Packers' Association was formed in 1885 for the mutual benefit and advantage of the members and for the intent and purpose of using all lawful means of enforcing laws enacted for the protection of the producers, dealers and consumers. The organization has had five name changes over the years and today is known as Associated New York State Food Processors, Inc. Prior to 1893, the Association had 26 active members, reaching a peak of 110 active members in 1957.[2] Later, wineries became eligible to participate in the organization and today there are over 200 winery members.

During the years prior to World War I, Clarence Birdseye, while working as a fur trader in Labrador, noticed that fish frozen quickly in the bitter cold

Cold frames of a truck farmer near Jamaica on Long Island in 1916. Courtesy of New York State Archives.

tasted like fresh when thawed and cooked. He applied this principle to vegetables in the 1930s. The Snider Packing Corporation of Albion utilized this knowledge in 1932 and began the production of frozen vegetables for a subsidiary of General Foods. In 1943, General Foods purchased Snider, which became known as the Birdseye Division of General Foods.[3]

The consumption of frozen fruits and vegetables skyrocketed from almost nothing in 1932 to many

---

2   The Associated New York State Food Processors, *A History of the Canning and Freezing Industry in New York State* (1985) p. 19-20

---

3   Ibid. p. 63-64

A 60-acre field of cauliflower at Hauppage on Long Island, in 1934. Courtesy of New York State Archives.

millions of boxes by the end of the 1940s. Today the Birdseye name continues to be important to New York vegetable and fruit producers, with the Birdseye brand still used today by producers of the Pro-Fac Cooperative.

In 1958, representatives of Curtice Brothers Company, the Burns-Alton Corporation and Haxton Foods, Inc. consulted with the management of the Cooperative Grange League Federation (GLF) to explore the possibilities of a joint venture of farmers and processors under control of the farmers. After thorough exploration and a number of meetings, the Pro-Fac Cooperative was formed in October, 1960. Subsequently over 500 fruit and vegetable growers purchased stock in the company in direct proportion to the amount of product they promised to deliver to Pro-Fac each year for the next three years.

Haxton Foods withdrew from discussions but Curtice Brothers and Burns-Alton continued talks and in 1961, Curtice-Burns, Inc. was established to process and sell products grown by the members of Pro-Fac. Shortly thereafter, in accordance with the master plan, Pro-Fac acquired the assets of Curtice-Burns and a year later Pro-Fac purchased Haxton Foods. Pro-Fac increased in size by acquiring facilities from other manufacturers and extended its growth beyond New York in 1967. Sales of the company increased from $13 million the first year to over $300 million in 1979.

In 1995, Curtice-Burns came off the public market and became a privately owned company. It later became known as Agra-link and then Birdseye. Currently, Birdseye is owned by a private equity company and Pro-Fac. Today, there are approximately 250 New York farmer members of Pro-Fac

A wagon-load of cabbage heading to the kraut factory, circa 1900. Courtesy of the OHA Museum & Research Center.

growing over 150,000 acres of vegetables, including sweet corn, beets, carrots, peas, string beans and cabbage.

There are several large vegetable processors, with plants in New York, including Allen Canning Co., with plants in Bergen and Oakfield; Seneca Foods with plants in Marion and Williamson; the Great Lakes Kraut Co. in Shortsville and Goya Foods Great Lakes of Angola, which processes a variety of beans. In addition to Pro-Fac growers there are other farmers, not members of Pro-Fac, growing vegetables for the processors.

Not all of the vegetables eaten during the winter were canned. The 1900 U.S. Census showed New York had over 13 million square feet of glass greenhouses, equaling about 325 acres. Some of this greenhouse area was used for non-vegetable items, but there were fresh vegetables available in the winter for the wealthy living in urban areas.

Looking at census figures for the 1800s we find a substantial amount of peas were grown. These, however, were not the peas we are familiar with today. They were a hard, gray pea, called the Canada field pea, and were grown as a feed for swine and other livestock.

The garden peas that we use today were also grown in the 1800s, but not in large quantities until late in the century when they were canned similarly to tomatoes and sweet corn. Pea vineries, which were little more than a roof to cover the machinery, were constructed in a number of farm communities.

A pea threshing station near Tully in Onondaga County in 1923. Courtesy of New York State Archives.

Farmers contracted a specific acreage and were required to mow the vines and haul the vines with peas to the vinery whenever the vinery's field representative determined they were ready. After mowing, the vines were put into small piles and pitched by hand onto wagons or trucks and drawn to the vinery where they were pitched to a conveyer. Machinery then shelled the peas and conveyed the vines to a stack. The shelled peas were transported to a central cannery for processing. During the following winter, the farmers returned to the vinery and pitched the vine silage on their vehicle to take back to the farm to feed their cows. The vine silage had a powerful odor but the cows seemed to relish it. For a number of years, specialized pea harvesters have shelled the peas in the field, along

TABLE 26 – Farms Growing Potatoes, Acres Grown and Yield
US and NY Census

Year	New York Farms Growing Potatoes	Acres of Potatoes (000s)	Average Yield in Bushels per Acre
1845*	NA	255	91
1875*	NA	358	102
1900	194,914	396	96
1925	140,610	285	136
1950	37,938	112	247
1997	544	24	460
2007	860	19	NA

*New York Census

Digging potatoes in 1896. Note the barrels for the potatoes, courtesy of the La Fayette Historical Society.

New York led the nation in the production of potatoes for about a century. In 1840, New York produced over 30 million bushels of potatoes with Maine, a distant second, producing slightly over 10 million bushels. New York's domination in potato production continued until 1920 when it was surpassed by Minnesota. **Table 26** shows the number of New York farms growing potatoes, the acreage and the yield per acre for several different periods of time.

There are several things that stand out in **Table 26.** The number of farms growing potatoes, in 1900, was 86% of the 226,720 farms in the State. The average acreage of potatoes on

the same principle as a corn combine, where they are loaded on a truck and transported to the factory for canning or freezing.

New York produces a number of additional vegetables, but in smaller acreages, including collards, endive, garlic, eggplant, herbs, peppers and radishes. An even greater array of vegetables can be found in the numerous green markets that have appeared in hundreds of communities around the State.

each farm was two acres. There were a number of reasons for so many farms growing potatoes. Every family ate potatoes, the families were larger with most of the family members performing a great deal of physical labor, and potatoes were a great source of calories at very little cost. Potatoes could also be stored for several months in the cellar of the house or in a root cellar, providing food for the family or

Potatoes in a field at Malone, circa 1920. Courtesy of New York State Archives.

for sale, during much of the year. New York also had a large population that provided a ready market for potatoes grown close to consumers. Another important reason why so many of the farms grew potatoes was that potatoes grew well in New York's climate and soils.

Today, less than 3% of New York farms grow potatoes, even though we still have more consumers and the climate and soils to produce good yields. When we look at the yield per acre, we see the great improvement that has occurred in the last century. This change is the result of improved varieties, increased fertilization and improved control of disease and insects. The machinery and knowledge necessary to produce potatoes in today's competitive market requires large-scale specialization. The relatively small amount of potatoes consumed by

today's farm family would seldom warrant more than a small portion of their garden. Today, in addition to farms producing a large acreage, there are some small farms producing potatoes for specialized markets that will pay a premium.

The author had the opportunity to grow up on a farm with a variety of livestock and crops, including potatoes, and his experiences were duplicated in proximity to every good sized city in the State, which makes the following experiences worthy of sharing.

"My father grew a few acres, as did my grandfather and great grandfather. We lived 20 miles from Syracuse, which provided a good market for potatoes. For over 50 years, during the fall of the year, my family transported potatoes by wagon or sleigh the 20 miles to the Syracuse public market. They would leave the

A truck farmer loading potatoes and cabbages near Jamaica on Long Island. in 1916. Courtesy of New York State Archives.

farm about 11:00 p.m. to arrive at the market when it opened at 5:00 a.m. and return back to the farm late in the afternoon. Thirty bushels was a big load and with a price of about 60 cents a bushel they grossed $18 for their work of growing those potatoes and marketing them.

Supermarkets were uncommon until the middle of the 20th century, but instead Syracuse had dozens of small corner groceries, whose proprietors also arrived at the 5:00 a.m. market opening to purchase the needs of their store for the day. In addition to the storekeepers there were hucksters, with horse and wagon, there when the market opened, to buy fruits and vegetables that they peddled from house to house. There were also a number of people who came to the market to buy produce for their families. In the fall, a family usually purchased their winter supply of potatoes, from 5 to 20 bushels, which we delivered to their homes. Normally the potatoes were carried from the street to the back door, down into the cellar, and then carried to the front of the cellar to be dumped into their potato bin. The author had the opportunity to experience all of these things except that my dad had a truck and it made the market day about 10 hours shorter."

Almost all of our farm neighbors had similar experiences from 1840 to 1940. The farmers that grew potatoes and lived further from a population center often sold their potatoes to a dealer who transported them to market by canals or railroads.

It is worth noting the movement of major potato production centers in both the US and State over the years. First place in US acreage of potatoes moved from New York to Minnesota, to Maine to Idaho. In New York, Washington County was first in 1875, Steuben first in 1900 with Steuben still first in 2007, followed by Wyoming, Wayne and Suffolk.

Dry beans have been grown in New York since the early 1800s. White beans were the most common in the 19th century with 16,231 acres of all beans grown in 1845. Red kidney beans became the predominant variety in the 20th century reaching 165,000 acres, on 11,406 farms in 1950 with a yield of 2 ¾ million bushels. The acreage has been decreasing since the mid-1900s, with 16,218 acres grown on 143 farms in 2007.

Chapter XI
Fruit, Nuts, Berries, Nursery, Sod, Woodland and Other Crops

In 2007, fruits, nuts and berries were produced on 3,227 New York farms providing over $363 million in sales. Wayne County had sales of $85 million in 2007 followed by Ulster with $42 million and Chautauqua with $41 million. New York State has been a long-time leader in orchard-grown products. For example, in 1840, the value of New York's orchard products was $1.7 million, more than twice that of the next state, Virginia, with its orchard products valued at a little over $700,000. In the 1850 census, New York's orchard products were valued at $1,761,950, with Oswego County first with $70,605 in value.

APPLES

New York is the second largest apple producing state in the United States with almost 50,000 acres of apple trees that produce over a billion pounds of apples. The predominating apple growing areas are along Lake Champlain, Lake Ontario and the Hudson River. There are 1,350 farms with a total of 49,966 acres. Wayne County is first with 20,862 acres, followed by Ulster, Orleans, Niagara, Clinton and Columbia. The 2007 New York Agriculture Statistics estimated the 2007 crop at 1.31 billion pounds with a value, if sold at packing plant prices, of $286 million.

TABLE 27 – New York Farms with Apples, Apple Tree Number or Acres, and Crop Value

Year	Number of Farms	Number of Apple Trees (000s)	Value ($000s)
1865	NA	9809	NA
1875	NA	18,279	NA
1900	NA	15,055	NA
1930	105,127	10,302	1,500
1940	61,540	6,581	9,000
1950	38,025	4,976	18,000
1997	1,557	8,030	141.300 (a)
2007	1,350	49,966*	286,000 (a)

*Acres of apple trees   (a) New York State Agricultural Statistics   NA – Not available

King Cider Mill on an Erie Canal feeder near Syracuse, circa 1905. Note the wooden barrels for the cider at the side and on the dock. Courtesy of Jamesville Community Museum.

The apple originated in the Caucasus Mountains of Asia and was carried by humans to the Middle East, Europe and eventually North America. In about 800BC, Homer referred to the apple in the Odyssey and in 79AD Pliny the Elder describes 20 varieties of apples. China is the world's leading producer growing about 35% of the world's apples, followed by the United States, where the state of Washington is the largest producer followed by New York.

Europeans brought the apple to New York where the Native Americans saw its value and planted orchards. Upstate New York settlers, in the late 18th century, found some of these orchards, since abandoned, and harvested the fruit. Few settlers had the enjoyment of a ready-made orchard but within a year of arrival planted apple seeds for trees of their own. **Table 27** lists the number of New York farms growing apples, number of apple trees or acres of apples and value of the apple crop.

In the 1800s, regardless of their location or the type of soil, almost every farm family in New York had an orchard with a few apple trees and usually a couple of pear, prune and cherry trees. The New York

Censuses of 1865 and 1875 do not provide information as to the number of farms with apple trees or of the value of the fruit from these trees, but it is interesting to note how the number of apple trees decreased so rapidly after 1875. Almost all of our farmlands had been settled by 1875 with appropriate orchards on most farms. Apple trees will produce fruit for many years so few young trees were being planted, other than on farms that planned to continue apples as an enterprise. As farms gradually consolidated, orchards on many farms were abandoned. I have included the figures for the value of the apple crop, in 1930, 1940 and 1950 to show how specialization in apple production that was beginning during those time periods, increased the value of the apples produced, while the number of trees decreased by half.

Apple cider was an important product from apples in colonial days and throughout the 1800s and early 1900s. Although it is still a desirable beverage today, the per capita consumption has substantially decreased. **Table 28** shows the number of bushels of apples harvested and the number of barrels of apple cider produced in three census years.

In 1875, New York's population was less than five million. With over a half-million barrels of cider produced it equated to a barrel for every nine people. If a barrel was 50 gallons there would have been over five gallons produced for each New York resident. It took about 16 bushels of apples to make a barrel of cider so about 8 million bushels of apples were pressed into cider in 1875. Most of the cider was stored for fermentation into hard cider. The poor quality of water in many localities created the need (or excuse) by some of New York's residents to drink hard cider rather than water. A temperance movement, as well as improved safety of drinking water, decreased the production of cider.

An 1826 advertisement by the Albany Nursery, owned by Jesse Buel, a renowned agriculturalist, and his partner James Wilson, lists 120 varieties of apple trees for sale. A few of the apple names are recognizable today, including pippins, sweetings and greenings, but most have been forgotten and replaced with improved varieties like the McIntosh, Empire, Delicious and Gala. Not only do the apples look and taste different, but most of today's apple orchards look entirely different than the ones of the past. Great strides have been made in the production of improved varieties of apples. An

TABLE 28 – Bushels of Apples and Barrels of Cider Produced
NY and US Census

Year	Bushels of Apples (000s)	Barrels of Cider
1855	13,689	273,639
1875	23,118	536,131
1900	24,111	145,953

Note: 1855 & 1875 New York Census, 1900 US Census

example is the use of the "gene gun" invented at Cornell University. It was used in 1989 to successfully transfer an anti-bacterial gene from a moth into a tree that was susceptible to fire blight, making the tree resistant to the blight. Today, when driving by a newly planted orchard, most people unacquainted with apple growing think they see a grape vineyard. These trellis apple orchards have from 600 to 2,500 apple trees to the acre, depending upon their spacing in the row, which may be as close as three feet while the distance between rows, is usually 12 feet.

There are different trellis arrangements, with the two most common being the single vertical and the "Y". The rootstock for the apple trees is disease resistant and is bred to produce a dwarf tree, rather

Trellis apple orchard in 2009. Many new orchards of this type are being planted in New York.

A Y-style trellis apple orchard in 2009.

An orchard of large apple trees at Livingston in Columbia county, circa 1925. Courtesy of New York State Archives.

than the huge, high trees of 50 years ago. These small trees, on a trellis, allow more sunlight on the fruit, giving better color and they also produce larger apples. At harvest time it is much easier to pick trellis apples.

Bees are essential for the pollination of apples and in recent years bee colonies have been disappearing without a trace. Fortunately there have continued to be enough bees for the successful pollination of apples. Apple growers rent bee colonies at blossoming time, normally a colony for every 3 to 4 acres of apples, from apiaries to assure successful pollination.

Apple growers have found spraying is necessary to control insects and disease in order to produce a high quality apple product, but are very careful to never spray when the bees are in the orchards. They also use an Integrated Pest Management (IPM) program to minimize the amount of spray materials used and the necessary times to spray. The planting of disease resistant varieties has also decreased the need to spray.

Apples are still picked by hand since successful apple harvesters have not been developed. Prior to the last 50 years, apples were harvested with mostly local labor. The industry was not as concentrated as it is today and apple picking often provided an opportunity for children and housewives to earn extra money. People from locations south of the US, who work here on a seasonal basis, now pick most of the apples.

Until after the 1850s, apples were consumed whole, as cider or dried. Many apples were shipped to metropolitan markets in the fall, but were only available as fresh apples, a few months of the year. Apples could be stored for several months in a root cellar, a covered depression in the ground where they wouldn't freeze, or in the cellar of the home. Since it was difficult to keep apples fresh, they were commonly sliced in the fall, and hung in the kitchen to dry. Dried apples were used when the fresh ones were depleted. When railroads came into general use, they provided faster transportation, permitting apples to be grown a greater distance from large markets. With improved transportation methods and the use of ice for refrigeration, New York apples began to be shipped to other countries. Whole apples were normally packed in wooden barrels for shipment, and dried apples were packed in both wooden boxes and barrels for shipping. Small canneries and commercial drying houses began to appear in many communities after the Civil War to provide applesauce, apple butter and dried apples.

Refrigerated storage of apples, mostly after 1900, extended the life of apples and permitted canneries and drying houses to operate with a longer season. The use of controlled atmosphere storage during the past 50 years, has given a big boost to apple consumption. In 1940, Dr. Robert Smock of Cornell University experimented with reducing oxygen and increasing carbon dioxide in apple storage facilities in. By selecting the most appropriate levels of each, as well as providing appropriate refrigeration and humidity, it is possible to maintain the fresh quality of apples for up to 10 months. The most desirable levels of oxygen and carbon dioxide vary with the variety of apple. The apples that are stored for fresh eating are usually held in controlled atmospheric storage, as well as apples stored for commercial processing. With the use of controlled atmospheric storage, processing plants are able to process apples 12 months of the year. The K.M. Davies Company, with a large storage facility in Williamson, installed their first controlled atmosphere storage in 1964.

In Wayne County, Dr. Peppers Snapple Group, Inc., operates an apple processing plant producing Mott's Brand applesauce and other apple products. It processes approximately 40,000 bushels of apples a day, many of which are grown in New York. Another processor is Cahoon Farms of Wolcott who slice and dice apples for both quick freezing and drying. At Lyndonville, in Orleans County, a company dries apples, blueberries, cherries and currents for shipping all over the world.

The Premier Apple Cooperative is a recent example of the successful cooperative marketing of an agricultural product. New York apple growers are working with growers in other states for the mutual benefit of all. In 2000, a New York Apple Industry Task Force was appointed to recommend changes that would improve the profitability of New York apple growers. The industry had gone through a horrible decade and most growers had experienced serious erosion of their net worth and many were facing bankruptcy. This task force of 24 people represented all phases of the apple industry with Commissioner of Agriculture & Markets Nathan Rudgers serving as honorary chairman. One of the group's recommendations was to form a grower cooperative that would allow apple marketers to work together to gain market power while being exempt from anti-trust laws under the Capper-Volstead Act of 1922. The primary objectives were to improve industry communications, the flow of market information, apple quality and the net income of apple growers.

Since the Capper-Volstead Act specified that the members of the cooperative must be "agricultural producers" an eleven person board was set up, with growers making up the majority. In order to include apple marketers, who were not growers, a Marketing Advisory Committee was established. Its purpose was to advise the Board on prices and other trade policies.

At the end of the first season it was determined that the organization could not accomplish its mission as a New York organization. Since all the major Eastern apple states had the same problems and the by-laws of the organization did not restrict membership in any way, members and marketers were recruited from Michigan, Minnesota, Ohio, Pennsylvania, Virginia, North Carolina, Massachusetts and Maine. The cooperative now represents over 75% of the wholesale Eastern fresh apple volume.

The first year, 2001, Premier had little effect on prices as the organization was just getting started and the committee members were learning to work together. In 2002, however, there was a short crop and the committee recommended significantly higher prices. The efforts of Premier that year resulted in Eastern fresh apple growers receiving $2.00 per box more, an increase of 16%. While

Cherry picker circa 1960. Courtesy of New York State Archives from New York State Ag & Mkts.

the results have varied each year, two independent econometric studies have shown that Premier Apple had increased returns to the growers by over $1.00 per box since its beginning.

Partially due to the efforts of Premier, high quality New York apples are prominently displayed year-round in grocery stores all over the world. Some of the countries purchasing New York apples are the United Kingdom, Canada, Israel and Costa Rica. Apples are also available in the stores, in a great variety of applesauce products and even sliced ready-to-eat. A few years ago McDonalds' Restaurants added fresh sliced apples to their menu, providing a big boost to the consumption of fresh apples.

There has been an increased interest in organic apples by the consuming public in recent years, but in 2009 Mott discontinued its organic line because of insufficient demand. The overall demand for New York apples has been increasing because of growing interest by consumers in better health and the availability of apples in more forms and in more places. Great quantities of applesauce and other apple products are now sold in single-serve containers. New York's 450 farmers' markets, along with numerous parking lot markets, provide increased opportunity for marketing of apples and other fruits. This increased demand has encouraged a large increase in the number of new orchards being planted.

CHERRIES

New York is and has been a large producer of both sweet and tart (sour) cherries. In 2007, there were 301 farms with 819 acres of sweet cherries and 201 farms with 2,041 acres of tart cherries. The census has provided data on both types of cherries since 1940 and tart cherry production generally was about six times as great as sweet cherry production until recent years. Until the latter part of the 1900s there were dozens of small cherry processors in the State. Currently there are only a few firms processing cherries. There were a great many tart cherry trees removed from production in the last two decades because production became unprofitable. Fresh sweet cherries are popular with the consumer and sell for a higher price, making their production more popular. The New York Census of 1865 lists 1,081 bushels of cherries with a value of $2900. There were certainly many more cherries produced than shown by the census, but the families producing them consumed the majority of those cherries.

GRAPES

The 2007 U.S. Census shows that 1,438 farms in New York are producing 42,544 acres of grapes. The leading county in grape production is Chautauqua with 547 farmers producing 22,276 acres of grapes. Although some of those grapes are used for wine, the majority is used to produce grape juice. Yates County is second in grape production with

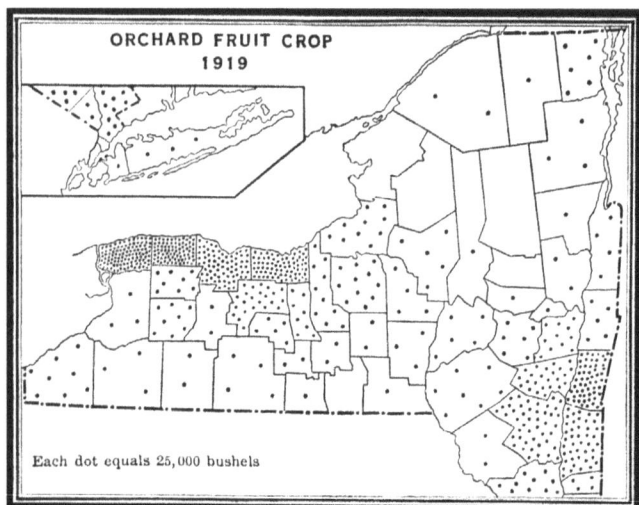

Each dot on the map indicates 25,000 bushels of orchard fruit grown in 1919. Courtesy of New York State Archives.

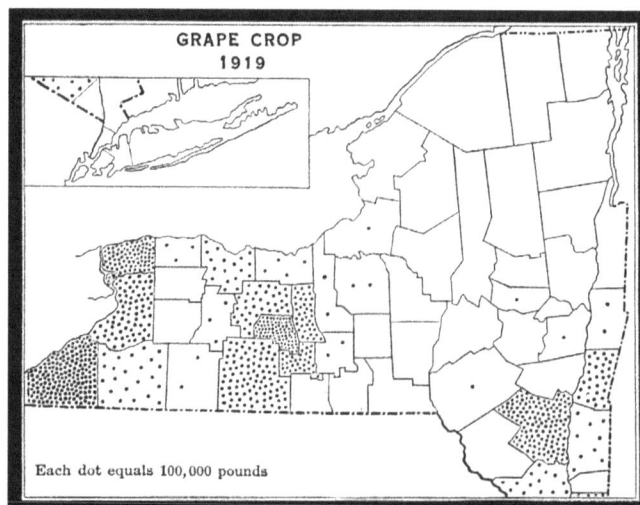

Each dot on the map indicates 100,000 pounds of grapes grown in 1919. Courtesy of New York State Archives.

Hillsides covered with grapevines, berry bushes and orchards at Marlboro in Ulster County, circa 1913. Courtesy of New York State Archives.

168 farms with 6,270 acres of grapes. The majority of those grapes are used in the production of wine. During the decade between 1997 and 2007, the number of grape growers in the State increased by 176 and the acreage of grapes by over 9,000. There has been a large increase in the number of wineries during the last 25 years, with the number well over 200 at this time.

New York's Department of Agriculture and Markets statistics for 2007 estimate that of the 160,000 tons of grapes produced in 2006, there were 4,000 tons of fresh grapes with 77% of the remainder processed for juice and 23% processed for wine. The total value of the 2007 grape crop is estimated at $49.2 million.

Almost every farm had a grapevine or two during the 1800s and early 1900s for their family's consumption. The 1930 Census shows over 38 million grape vines in the State with the number decreasing for some years after 1930 but increasing in the last decade or two. (The grape census numbers are now given in acres rather than number of vines.) In 1865, the State produced 69,960 gallons of wine led by Monroe County, which produced over 12,000 gallons, followed by Ontario and Steuben each producing over 5,000 gallons.

Currently, the National Grape Cooperative Association, Inc., located in the village of Westfield in Chautauqua County, has 344 New York members growing over 12,000 acres of Concord and Niagara grapes. These members, along with 900 grape-growing members from other states and Canada, own Welch Foods, Inc. Welch processes and markets the grapes, in a variety of products, throughout the world. Sales of products from this plant, along with other Welch plants, total $700 million annually. Although there have been challenging times in the grape industry, as in other agricultural industries, the volume of sales and profitability of Welch's have been increasing since 2007. Grape processing in Westfield goes back to 1893 when Thomas Bramwell Welch constructed a grape juice processing plant, which was sold to the National Grape Cooperative in the 1950s.

All of the juice grapes are harvested mechanically, a tremendous savings of labor compared to the hand harvesting methods used until the 1960s. National Grape has adopted a program of sustainable viticulture. It utilizes practices that have minimal impact upon the environment, produces safe, high quality products and provides a work environment that encourages personal employee growth.

PEACHES

Peaches are very much enjoyed but difficult to grow in much of New York because of the cold weather. Peaches have been grown since colonial times. In the report of Sullivan's 1779 campaign against the Iroquois, it stated he destroyed an orchard of 1,500 peach trees at what is now the village of Aurora, on the east side of Cayuga Lake. Wine can be made from a variety of fruits and peach wine used to be considered the very best. It is from peach wine that

Spraying a peach orchard in Livingston County in 1913. Courtesy of New York State Archives.

107

our expression "it's peachy" came from, meaning the very best.

The 1930 U.S. Census lists almost 2½ million peach trees in New York on 12,683 farms. The 2007 Census figures show 375 acres for that year, which would be far fewer trees than in 1930. Peaches still produce significant income, however, as New York Statistics estimate the 2007 crop at 6,300 tons with a value of $4 million. They also indicate the crop was 10% less than in 2006, which is not surprising because the size of the peach crop, like so many other agricultural crops, varies with the weather. If weather is favorable all year the peach crop is bountiful, but if there is a frost at blossom time, a dry year or a hailstorm, the crop can be a disaster.

PEARS

Pears were often found in farmer's orchards during the 1800s and early 1900s. In 1865, there were 6,600 bushels produced with a value of $16,000. Genesee and Rockland counties were the largest producers. The 1930 Census shows about 2½ million pear trees, but numbers grown gradually decreased since then to 1,510 acres produced on 375 farms in 2007. New York Agriculture Statistics estimate 2007 production at 11,000 tons with a value of a little over $5 million. Pears, like apples, can also be held successfully in controlled atmospheric storage.

PLUMS AND PRUNES

Plums and prunes are really the same fruit but because of the perception in the United States of prunes being a laxative or health food the name was changed to plum in 2000. In some parts of the world there is no laxative connotation and prunes are still prunes. Prunes were never as popular in New York as pears, peaches, apples and grapes, but there was at least one prune tree in the orchard of most farms a century ago. There were almost a million trees listed in the 1900 Census but the number has decreased to the point that there were only 367 acres grown on 208 farms in 2007.

APRICOTS AND NECTARINES

The 2007 U.S. Census also shows 98 acres of apricots and 117 acres of nectarines grown in New York. There are 25,000 apricot trees mentioned in the 1900 Census and in that census nectarines are included with peaches. Both apricots and nectarines have been minor fruit crops throughout New York's agricultural history.

BERRIES

The 2007 U.S. Census lists 1,254 New York farms growing berries on 4,314 acres. Ulster County had 299 acres followed by Onondaga with 252. Strawberries were the leading berries with 1,659 acres grown on 625 farms, while 621 acres of raspberries were grown on 525 farms. Onondaga County ranked first in both strawberries and raspberries. **Table 29** gives the acreage of several varieties of berries in different years.

With an increased interest by consumers in locally grown food products, organic food and green markets, we see an increase in the production of all berries between 1997 and 2007. Several notable changes are the increase in blackberry and dewberry acreage, the decrease in the acreage of currants and the increase, during the last ten-year period, in the production of raspberries. The most

TABLE 29 – Acres of New York Berry Production
US Census

Year	Blackberries & Dewberries	Blueberries	Currents	Raspberries	Strawberries
1900	2,060	NA	2,594	12,376*	7,311
1950	862	173	1,340	4,574	3,631
1997	95	722	3	378	1,617
2007	355	1,317	44	621	1,659

*Loganberries were included in this number

Picking strawberries near Middle Hope in Orange County, circa 1913. Courtesy of New York State Archives.

A field of grapes, currants and berries at New Hope in Orange County, circa 1913. Courtesy of New York State Archives.

noteworthy change, during the past 50 years, has been the increase in the production of blueberries.

At various times in our history one or two counties have been dominant in the production of an agricultural product. In 1900, Ulster County grew 1,240 acres of currants; almost half of the State's production and a sixth of the total value of New York's small fruits.. The same year Wayne County grew 2,500 acres of raspberries, about 1/5 of the State's raspberry crop.

The information provided in the Census numbers varies from census to census. I was intrigued to read that the production of raspberries and strawberries, as reported in the New York 1865 Census, was 2,565 bushels and 11,939 bushels respectively. The production of most farm products was measured in bushels at that time, so they apparently decided to use the same measurement for berries. If one were to fill a bushel basket with either raspberries of strawberries it would be interesting to see what the berries looked like in the bottom of the basket!

Many farms have initiated pick-your-own operations during the last 50 years. Consumers enjoy coming to the farm to pick their own berries, sometimes to save money but more often to enjoy the on-the-farm experience of harvesting berries for their family's use. Large supermarkets provide the majority of berries for New York consumers, selling berries from all over the world all during the year. Some of these supermarkets make a point of offering berries

from local farmers to their customers when they are able to purchase them. This has also helped to increase berry sales from local farms in recent years. A decade ago, a large cranberry farm was developed in Oswego County selling a number of acres of cranberries. Since the 2007 Census shows only one acre of cranberries in New York, the same acreage as in 1950, it probably was not successful.

NUTS

Nut trees provided by nature have been enjoyed by generations of New Yorkers. Hickory nuts, commonly called walnuts, and butternuts have made cakes special and provided an enjoyable evening treat for many families. Their use has decreased since other nuts, with easier access, became readily available in the supermarkets. The American chestnut was a common native tree ranging from Maine to Georgia until blight killed almost all of the trees in the early 1900s. They were easily picked off the forest floor in the fall and were enjoyed by thousands of New Yorkers. The majority of the chestnuts we see roasted and sold on the street corners of large cities today are of a different variety and come from Europe.

Commercial production of nuts is only a very small agricultural enterprise in New York. The 2007 U.S. Census indicates that there are 135 farms growing 377 acres of nuts. The varieties are Chinese chestnuts, hazelnuts, pecans and English walnuts.

TABLE 30 – Farms, Greenhouse Space, Acres and Value of Sales of Certain Agricultural Products
2007 US Census

Crop	Under Glass		In the Open		Value of Sales (000s)
	Number of Farms	Square feet (000s)	Number of Farms	Acres	
Aquatic plants	14	12	12	24	2,591
Cuttings, seedlings	39	431	4	2	4,446
Floriculture crops	1,070	25,430	584	1,597	225,918
Greenhouse vegetables & herbs	201	3,039	X	X	17,706
Mushrooms	12	69,517	X	X	238
Nursery stock	84	906	646	14,638	101,095
Sod harvested	X	X	14	6,868	33,260
Vegetable transplants	98	469	14	14	1,365

X: Nothing in this category

NURSERY, GREENHOUSE, FLORICULTURE AND SOD

The sales from nursery, greenhouse, floriculture and sod production in New York were $389 million in 2007, greater than the other groups of agricultural products produced in our State, with the exception of milk and dairy products from cows. There were 2,009 farms in New York producing these crops representing millions of square feet of greenhouses and thousands of acres of land. Suffolk County is by far the largest producer with over half of the sales.

The U.S. Census of 1840 lists the value of nursery produce sold at $75,980. This seems like a small sum today and it is, compared to current sales. It did, however, provide for the nursery needs of a relatively small urban population in a largely agricultural state. The nursery aspect of New York's agriculture was largely ignored in early census information because it was a fledgling part of a very large agricultural industry.

The 1900 U.S. Census does show us that we had a growing New York industry with 1,496 acres of flowers and plants providing sales of almost $3 million. The nursery industry was also thriving with over 8,000 acres and sales of $1.6 million. At that time Nassau County was first in the production of flowers and plants, with Monroe County first in the production of nursery products. By 1950, the sales of flowers and plants surpassed $23 million and the sales of nursery products exceeded $34 million.

The breadth of the nursery, greenhouse, floriculture and sod industry is so great that the 2007 U.S. Census broke it down showing size of activity in several areas. **Table 30** provides some of that information.

There are several key points in **Table 30**. The production of floriculture crops is not only a large industry with over $200 million in sales but it also has over 1,000 farms involved in its production. The farms growing these products have over 100 million square feet of greenhouses, which is the equivalent of over 2,000 acres. The investment in greenhouses represents many millions of dollars. Open fields, of course, represented the largest area with over 22,000 acres.

Suffolk County was the leader in sales of floriculture crops, nursery crops and sod by a wide margin. Orange County ranked second in the production of floriculture crops followed by Erie County.

WOODLAND PRODUCTS

The woodlands of New York are an extremely important resource. They are not only a source of fuel, lumber and a variety of wood products, but are also critical to our environment by removing carbon dioxide from the air while producing oxygen. They also improve our climate by holding moisture in the soils, aesthetically improving our environment, providing a friendly habitat to wildlife and enhancing our lives in numerous other ways.

Upon the arrival of the white man to what is now New York some 400 years ago, our State was almost entirely woodland. Settlement, agricultural uses and open space for the needs of an increasing number of residents, have substantially diminished our woodlands over the years. Many acres of land were cleared in the 1800s that would have been better left as forest. During this past century, however, large quantities of previously cleared land have reverted to woodland. In 1910, the percent of New York's land area in farm woodland, unimproved pasture and parks was 47.5. In 1992, the percent of New York's land area in farm woodland, mostly wooded parks, government and privately owned woodland reached 70.7.[1]

New York State increased its ownership of land by both reforestation and purchase for parks and recreation during the 1900s. During the period 1910 to 1992, the acreage of farmland decreased by about 65%, equaling over 14 million acres, while mostly forested land, consisting of parks and privately owned land other than farms, increased by almost 11 million acres.[2]

There were a little over 1½ million acres of farm woodlands listed in the 2007 U.S. Census. Steuben County ranked first with 96,000 acres followed by St. Lawrence, Chautauqua and Cattaraugus counties. A number of other counties have a greater area of forest but are located where there are large acreages of government owned land.

Our woodlands were critical to the prosperity of the majority of our farms until the middle 1900s. Few people would recognize their importance other than our older generation, who lived on a New York farm. Many winter days, between caring for the animals in the morning and the evening, were spent in the farm's woodlot or in and around the farmyard harvesting the wood products needed on the farm for the following year. The major item was wood to furnish fuel for heating and cooking. The wood was required for the main home and perhaps another home for a son or a hired man working on the farm. Seven-foot sections of tree trunks were split into fence posts; wood was used to heat water for butchering the hogs and to heat the farm shop where tool handles, eveners, wagon tongues and ox yokes were fashioned. Logs needed to be cut and sawed into lumber to make wagons, repair barns and for construction of new buildings.

---

1   Stanton, Bernard F. & Bills, Nelson L. *The Return of Agricultural Lands to Forest* Department of Agricultural, Resource & Managerial Economics, CALS Cornell p. 8-11

2   Ibid. p. 12

TABLE 31 – Forest Products from New York Farm Woodlots
1950 US Census

Item	Number of Farms	Number
Firewood	40,870	947,823 cords
Fence posts	21,114	4,182,671
Saw logs	9,092	52,914,000 board feet
Pulp wood	1,013	33,152 cords
Piling & poles	374	53,190
Standing timber sold	1,981	$1,017,691
Christmas trees & miscellaneous	1,053	$144,796
Maple syrup & sugar	7,097	$1,001,238

A much greater number of farmers produced items from the farm woodlot in the 1800s, but unfortunately this data was not collected at that time. The 1840 U.S. Census did report, however, that farmers sold over a million cords of wood that year, which was in addition to what they used on their own farms. In that same 1840 census, Virginia ranked second with less than half the number of cords of wood sold.

Table 31, with data from the 1950 U.S. Census at a time when the utilization of wood products on the farm from the woodlot was decreasing, provides a snapshot of the woodlot's uses. The cords of wood referred to in the table are full cords, which measure 4 feet by 4 feet by 8 feet. The total value of all of the forest products produced in 1949 was almost 4 million dollars.

The author's memories of the production of forest products on the farm are vivid because he spent many days helping his dad during the late 1930s and the 1940s. A hearty breakfast was enjoyed after milking the cows. Then it was time to harness the horses, hitching them to the bobs (similar to a wagon but with four runners instead of four wheels), cleaning the stables, spreading the previous day's fertilizer produced by the cows and horses on the field where next year's corn would be grown and loading the bobs with the simple tools needed in the woodlot. Arriving in the woodlot we viewed a scene of beauty. The wind may have been blowing outside the protective barrier of trees but inside tranquility prevailed. The forest floor was covered with a clean layer of snow interrupted only by a few rabbit tracks. Majestic trees poked their heads into the sky. A tree was selected, a senior citizen of the forest that had passed its prime, the crosscut saw was used to make an undercut on the side selected for the tree to fall, an axe made large chips of wood fly from above the undercut and with a man on each end of the crosscut saw, the music of the saw, powered by the muscle of man, brought the tree to the ground with a thundering crash.

When producing firewood, the trunk of the tree was cut into four-foot lengths and these lengths were split with iron wedges and mall into pieces that a man was capable of lifting. Branches too small for firewood were removed with an axe, and branches large enough for fuel were cut into lengths of up to 12 feet. As dinnertime approached, the bobs were loaded with wood and we left the tranquility of the woodlot, heading for the homestead where dinner awaited both man and horse. After a hearty meal, the wood was stacked vertically in a pile, adjacent to loads from previous days, awaiting the day when the buzz saw would turn the wood into appropriate lengths for burning. After the buzz saw did its work, the chunks of larger diameter awaited man and axe to turn them into smaller pieces appropriate for the housewife to feed her kitchen stove for cooking and heating. Late in the summer, when the wood had dried, it was hauled to the woodshed for use in the

TABLE 32 – Maple Sugar and Maple Syrup Production in New York
US Census

Year	Pounds of Maple Sugar in (000s)	Gallons of Maple Syrup	Number of Trees Tapped
1850	10,357	NA	NA
1860	10,816	131,843	NA
1880	10,693	266,390	NA
1900	3,623	413,159	NA
1940	131	800,410	3,146,970
1950	36	399,646	2,114,000
1997	NA	308,692	1,881,618*
2007	NA	229,486	1,342,165*

*Number of taps. Large trees have multiple taps.

Boiling sap for maple syrup, circa 1905. Courtesy of Ralph Bratt.

kitchen stove and fireplace or to the cellar if the farmhouse had a furnace. If there was a young boy in the home, he was given the task of keeping the wood-box filled on a daily basis. The wood not only heated man when it was burned but also heated him a number of times in preparing it for burning!

These experiences were typical of those on thousands of other farms scattered throughout the State for over 100 years. Some days logs were cut for lumber and hauled to the sawmill, while other days fence posts might be cut from rot-resistant red cedar or locust trees. Some farms used a drag saw to cut the trees into short lengths and eliminating some of the splitting with wedge and mall. All across the State the woodlot was treated as an important farm crop that needed to be cared for and harvested when ready, because much of the livelihood of the farm family depended upon it.

Christmas tree production was not a prominent agricultural industry until the 1900s. Many farms set trees during the mid-1900s for sale as Christmas trees. The advent of the artificial Christmas tree has slowly diminished the market for fresh trees during the last 25 years. In 1997, there were 1,648 farms with 32,599 acres of Christmas trees, but by 2007 the number of farms decreased to 1,154 with 20,267 acres. The value of Christmas trees, along with other short rotation, woody crops, was listed as a little less than $9 million in the 2007 Census. In that census, Monroe County was listed as selling the most, with 27,251 trees, and Steuben County second with 21,217.

The maple tree was important to the Native Americans for its sweet, tasty flow of sap. The white man learned of its value, producing mostly maple sugar at first, and later maple syrup, called molasses. **Table 32** provides information regarding the production of these maple products.

It is hard to imagine over 10 million pounds of maple sugar produced each year in the middle 1800s. A pound of maple sugar represented approximately 15 gallons of sap, with the water boiled off into steam to furnish the sweet end product. A gallon of maple syrup, called sugar molasses in early times, required close to 40 gallons of maple sap. Great quantities of wood were required to boil the water from the sap and to make the delicious maple syrup and maple sugar. Cane sugar, which was grown extensively in the Caribbean, was available in the villages and cities, but was costly. The

113

Harvesting hops near Cooperstown, circa 1900. Courtesy of the New York State Historical Association.

farmer could tap his maple trees, boil the sap and have maple sugar at no cost, other than his labor. The maple sugar ably filled the needs for a sweetener in their homes.

The production of maple sugar almost completely disappeared over 100 years ago with the advent of safe and convenient storage containers for maple syrup. Maple syrup production has been decreasing since 1940 because of the tremendous amount of labor required in its production relative to the price the producer is able to receive. New York has many maple trees that could be tapped, but it is unlikely that maple syrup production will increase. In 2007, there were 1,313 farms producing maple syrup. Lewis County farmers produced the most with 28,786 gallons followed by Wyoming with 26,016 and St. Lawrence with 20,575.

In the later 1800s and into the 1900s, the manufacture of willow baskets was a large home industry in Onondaga County. The 1900 Census lists 366 acres of willows with Onondaga County first, followed by Wayne County. Originally, wild willows were harvested for the baskets. Later they were planted in fields and furnished an annual crop for up to 20 years, before it was necessary to replace them with a new planting of willows.

Today, scientists are experimenting with the use of hybrid willows to produce ethanol. The New York State College of Forestry at Syracuse has been conducting research for a number of years to develop varieties that might be successful. Some of these willows are being grown south of Syracuse, but the current acreage is small.

OTHER NEW YORK CROPS

There are a number of other crops that have been grown or are being grown in New York. For many years, with limited availability of transportation, almost everything that was needed was grown in the local community. As transportation improved, agricultural products were transported greater and greater distances until today agricultural products come from all over the world. Improved transportation has also led to specialization and the production of crops in the most desirable climate, with the most appropriate soils, and where the costs of production are low enough to overcome the cost of transportation.

HOPS

From the 1840s to 1900, New York was by far the largest producer of hops, reaching production of over 17 million pounds on 27,532 acres in 1900. Otsego County was the largest producer with the adjacent counties producing most of the remainder. By 1920, the production had dropped by over 95% with only a thousand acres grown. The production of hops moved to other places and by 1960 there were only 14 acres grown in New York. Growing hops was very labor-intensive, especially at harvest time. During the peak production years of the late 1800s and early 1900s picking hops provided extra money for thousands of men, women and children from the surrounding area. The hops were dried in special barns with wood-fired heaters. A few of these barns still remain in Otsego, Oneida and Madison counties. Production moved gradually to the Western States and with the arrival of prohibition, New York hop production quickly came to an end.

FLAX

Flax was the source of linen, which was spun and woven from the long fibers in the flax stem. It was used in the home to make many of the clothes worn by the people of New York throughout the colonial period and into the late 1800s. Wool was the other major material for clothing at that time, and linen and wool were often combined to make a cloth called linsey-woolsey. The shorter flax fibers, called tow, were used to make sacks for holding grain, potatoes and other farm products. Another important product from flax was the oil from the seeds, called linseed oil. The 1825 New York Census lists 121 oil mills in the state making linseed oil from the flax seeds. Linseed oil was used in paint, ink and varnish.

The 1860 U.S. Census shows production of 1½ million pounds of flaxseed, with Rensselaer County producing almost half a million pounds and Washington County a close second. The quantity of flax fiber is not stated but it certainly would have made a lot of cloth. Cotton was being shipped north from the cotton producing states of the South, gradually causing a decrease in the amount of flax grown in New York. During the Civil War, when cotton was not available from the South, flax production increased, but after the war production rapidly decreased and by 1900, only 159 acres of flax was grown in New York.

SILK

Another New York crop of the past is silk. In the early 1800s, there was a silk production craze. Many people thought they were going to get rich producing silk. The only ones gaining wealth were the ones selling the mulberry trees that furnished food for the silk worms. Connecticut was a fairly large producer of silk with a production of over 17,000 pounds of silk cocoons in 1840. At this same time, New York produced 1,735 pounds. By 1860 the craze was about over with only 259 pounds of silk cocoons produced.

TOBACCO

Native Americans grew tobacco for many years in New York and small amounts were produced by white settlers in Colonial New York. The 1840 U.S. Census indicates that 744 pounds was produced in New York that year. Tobacco production started developing in the Town of Marcellus, in Onondaga County, around 1845 and by 1850 production in the State reached 83,000 pounds with over 73,000 of that produced in Onondaga County. Tobacco production was profitable, at that time, and by 1860 production surpassed 5 million pounds. Production continued to increase with 11,307 acres yielding almost 14 million pounds by 1900. With almost all agricultural crops, there are years when prices are good mixed with years of poor prices, and tobacco was a crop with extreme price changes. This, coupled with the fact that other areas of the country could produce tobacco more economically, led to decreasing production in the early 1900s. By 1925, production had dropped to 1,897 acres; 25 years later only 397 acres were grown. Driving through northwestern Onondaga County, buildings still remain that were once used to hang the lathe, with their stalks of tobacco for drying.

TEASELS

Teasels were a specialty crop produced in Skaneateles and several other towns in the western part of Onondaga County for over a century, beginning in 1833. Teasels were used in the manufacture of wool to raise its nap. The teasel industry, like hops and tobacco, was confined to a relatively small area of the State but was a crop of significant importance to the people in the area where it was grown. It provided work for hundreds of people during the fall harvest period. Mechanical nappers, made

Sorting and clipping teasels, circa 1910. Teasels were grown in the Skaneateles area and sold for use on machines that raised fibers on a cloth surface to form a nap. Courtesy of the Skaneateles Historical Society.

from metal, gradually replaced the teasel in the early 20th century. Teasels are listed in the 1920 U.S. Census with a crop of 78 acres producing 48,420 pounds. By 1950, production had decreased to 6 acres producing 6,400 pounds. Production was greater during the late 1800s but census information regarding the amount was not readily available.

SUGAR BEETS
In 1858, the first sugar was produced from beets in Europe. Sugar beet production came to New York in the late 1800s. A factory was constructed at Rome, New York to process the beets into sugar, but ended in bankruptcy in 1900. Factory owners were unable to convince local farmers to grow the sugar beets so beets had to be purchased from farmers in Western New York. The cost of shipping made the process too expensive. Sugar beets came to Central New York again in 1965, when a large sugar beet processing plant was constructed by PepsiCo, adjacent to the Seneca River at Montezuma, in Cayuga

County. The first year, 668 farmers harvested 14,436 acres of sugar beets they had contracted to grow for the company. Neither yields nor profits met expectations, resulting in the plant closing in 1969. During the five years the plant operated, farmers from 18 counties raised beets for the plant. In 1971, PepsiCo sold the plant, which was then operated for a few years by Clinton Corn Processing Co. for the manufacture of high fructose syrup. Sugar beet production then again left New York.[3]

GINSENG
For 3,000 years the Chinese have used ginseng as a medicine. Ginseng grows in the shade of hardwood forests, usually under maple trees in the Northeast. The roots sell for a very high price but are scarce and difficult to find. Native Americans gathered the roots for sale to traders in the 18th century and some of the white settlers also gathered them. George

---

3   Colman, Gould P. & Stockdale, Jerry D. *New York's Sugar Beet Fiasco*

Stanton of Apulia Station, in Onondaga County, began cultivation of ginseng in the 1880s and was able to grow a crop successfully. He served as the first president of the New York Ginseng Association and became one of the largest shippers of seeds and roots in the country. Knapp Brothers and Sons owned Ideal Ginseng Gardens in the same general area and had 1½ acres of ginseng growing under lathe for shade. Blight ended New York's ginseng industry in 1906.

BROOM CORN

During the period when the majority of necessities were made close to home, broom corn was grown in New York. Broom corn is not mentioned often in census information but the 1900 U.S. Census shows 356 acres grown with a weight of the tassels, which were used to make the brooms, of over 100 tons.

SWEET POTATOES

Most of us don't think of sweet potatoes as a New York crop, but sweet potatoes were grown, to a limited extent, during the 1800s. The 1850 U.S. Census lists 5,629 bushels grown, with 2,229 bushels in Albany County and 1,411 in Richmond County. In 1900, there were 8,681 bushels grown, with Suffolk County producing over 6,000 bushels. They were probably included in census information because sweet potatoes are a large crop in the South and the same census questions are asked in all of the states.

AQUACULTURE

The production of fish as a farm crop in New York has been limited because of our cold winters and the lack of suitable ponds to grow them. In the 1980s and 90s, methods had been developed to grow fish inside buildings in large tanks. So a few farmers with empty dairy barns decided to convert the barns to aquaculture. There was hope that with the large number of consumers in and close to New York a market might develop. Aquaculture farmers usually found that the cost of energy for heat in the winter and the cost of maintaining the equipment left no profit. There were 157 aquaculture farmers in 2002 with a total of $15 million in sales. In 2007, the sales had increased to $20 million, while the number of producers decreased to 127.

GRASS, CLOVER and ALFALFA SEEDS

In the days of the self-sustaining community, almost every farmer produced the grass seed needed for his farm or secured some from a nearby neighbor. The 1860 U.S. Census shows that there were 81,625 bushels of grass seed produced that year, with Jefferson County producing 9,523 bushels followed by Sullivan County. That same year there were 106,934 bushels of clover seed produced with Albany County growing 11,676 bushels, followed by Schoharie County with 10,044. By the 1900s, the production of both grass seed and clover seed had decreased. Alfalfa began to replace clover and timothy in the early 1900s with 15,000 pounds of seed harvested in 1930, increasing to 276,600 pounds in 1940 and dropping to less than half of that in 1950. Birdsfoot trefoil became a desirable pasture grass in the mid-1900s, with a little over 100,000 pounds of seed produced in 1950. The US 2007 Census shows 53 New York farms producing grass seed on 2,420 acres and 35 farms producing red clover seed on 1,675 acres. There were only 27 acres of alfalfa seed production on three farms in 2007.

Chapter XII
Barns, Fences, Drainage, Fertilization and Chemicals

Driving through New York State, a person sees a great variety of barns of many shapes, sizes and uses. Each barn has a story to tell about the history of the farm on where it rests. There are round barns, stone barns, steel barns, little barns and huge barns. New York has a wide variation in its climate and soils creating diverse environments and a wide variety of agricultural products. These differences provide us with an assortment of barn types designed to meet the farm's needs.

Until the first sawmill was constructed in a community, normally within a few years after the first settler arrived, buildings were constructed of logs or were simple structures made from poles covered with bark. In some instances the logs were squared with a broad axe, but usually left round for more rapid construction. Most settlers in a new community came from areas where there were sawmills and where frame construction was common. As agriculture expanded in a community, the need for frame barns grew. The erection of a sawmill made their construction possible.

It was not possible for one or two men to construct a barn by themselves, with the simple tools available at that time, so barn-raisings involving a number of nearby families were common. The family, living where the barn was to be constructed, cut the logs and hired a master carpenter to hew the beams, cut the mortise and tenons and prepare the barn for assembly. A day would be selected for the barn-raising and neighbors would come to help raise and assemble the barn. Pike poles and ropes were used to pull up bents and then beams were connected and braces attached. It required many hands to assemble the heavy sections. The farm owner would have a good supply of drink available and the farm wife, with the help of her neighbors, would prepare a feast for everyone. It was a hard day's work but also one of fun and festivity. In the following days, the carpenter and farmer would finish the barn now that the difficult part was completed. On occasion, after the rafters were in place, one of the gymnastically talented workers would stand on his hands at the very top of the barn.

A barn was usually the first frame building constructed on a farm. Shelter for the livestock and storage for the crops was a higher priority than a fine home. The frame home was constructed after the necessary farm buildings were in place. The original log home was then utilized for animals or storage until deterioration required its removal.

The Dutch were the earliest white settlers in New York and they constructed a distinctive heavily-timbered barn that was box-like in shape featuring a steep gable roof and double wagon doors, centered on each end with animal doors in the corners. This type of barn, now called the "Dutch" barn, was designed for a region dominated by grain farming. It contained an H-shaped structural frame with three to five large anchor beams, 20-30 feet in length, that supported sapling poles upon which bundles of grain were stored. There was a plank floor underneath for threshing grain and side aisles for cows and draft animals. These barns were constructed between 1630 and 1825 and were located in the Hudson, Mohawk and Schoharie valleys as well as in Central and Northern New Jersey. The Dutch Barn Preservation Society was organized in 1985 to encourage preservation of these barns, of which a few hundred are estimated to have survived.

The most common barn constructed by early English settlers in Eastern New York and later throughout Upstate New York, was the English threshing barn. It was usually 30 x 40 feet and consisted of three bays: storage for un-threshed grain on one end, a threshing floor in the middle bay and the third bay to house cows, draft animals and a granary for the threshed grain. There was a mow over the third bay to store hay for the livestock housed underneath. The sills for the barn frame were often set on large stones, which served as the foundation for the barn. Many times they did not build a floor where the un-threshed grain was stored, but a good plank floor was necessary for the threshing floor. Although thousands and thousands of these barns were constructed and still dot our countryside, many of

them have been modified, frequently incorporated into a larger barn to meet the changing needs of the farm. Often the original barn was extended on an end or a side and was raised to put a basement underneath to house dairy cows. Silos and milk houses were also added, totally changing the barn's original three-bay configuration.

Several small auxiliary barns came into use as our farms were settled. These were special-use barns that were generally smaller. The corncrib, for storing ear corn, and the pig house were two of the first buildings needed. Later, since pigs were a large consumer of corn, the two buildings were often combined into one. As the farm prospered, the family ceased their daily call behind a tree in the adjacent woods, and built an outhouse within a reasonable distance of the home, but not close enough to have the odors mix with those of meal preparation.

Chickens ran loose until the community developed sufficiently to provide a market for eggs produced beyond the family's needs. At that point, a simple building with nests was built to make egg gathering easier and also to protect the chickens from predators. The chickens were allowed to run outside

The author, with his father, ready to haul grain from a combine with the wagon in 1937. Small combines did not become common in New York until the 1940s. Barns in the background, from left to right, are: gambrel roof cow and horse barn, English threshing barn, tobacco shed, combination corn crib and hog barn and chicken house. Although this was a 90-acre farm, the number of specialty barns was typical of thousands of New York State farms.

An English threshing barn, 30 feet by 40 feet located in Onondaga County and constructed circa 1840. Thousands of these barns were built in New York during the middle of the 19th century.

during the day to eat grass and bugs, which were an important part of their diet. As some farms expanded their chicken production, larger barns for the chickens were constructed, and often a large dairy barn was converted into a multi-floor chicken house.

A growing farm business required a variety of barns to meet the needs of the various crops and livestock. Often a separate barn would be built to house the horses, wagons and carriages. Farms with large dairies had a separate barn for their cattle. Some farms needed a barn for their sheep and as new labor-saving equipment came to the farm, a tool shed was constructed to house this equipment.

As production of cheese and butter moved from the farm to the cheese factory and creamery in the last half of the 1800s, two additional farm structures became common. One was the icehouse, necessary for storing the ice used to cool the evening milking, and the other was the milk house where the milk was cooled and the milk cans stored.

Special barns were needed for special crops that are no longer produced in New York. Hop barns were

constructed in South-Central New York for drying and forming the dried hops into bales. There were also special barns for the storage and processing of teasels in the Skaneateles area. With the 1850 advent of tobacco production in Central New York, a new type of barn was required to hang the tobacco for curing. These barns almost always had a gable roof and were usually about 24-feet-wide by whatever length the farmer needed to store his tobacco, usually at least 48-feet-long. A stripping room was included at one end of the barn for removing the leaves from the tobacco stalks. Occasionally, part of an existing barn was modified to hang the tobacco. In 1907, the author's grandfather built a multipurpose gambrel roof barn that was used for potato storage, heifers, horses, grain storage, hay storage and tobacco. By purchasing an additional 24 acres a few years earlier, he lacked sufficient room in his other barns.

Initially, maple syrup production was accomplished by boiling maple sap in a large iron kettle in the woods and making the maple sugar in the home. In the latter part of the 19th century and continuing today, farms with a large sugar bush constructed a

Filling silo in a barn in Oneida County in 1912. Courtesy of New York State Archives.

special barn, called a "sap house", with an evaporator inside for producing maple syrup.

In the late 1800s, farms that were increasing their dairy production, built wooden silos for the storage of corn silage. In 1882, there were only 92 farmers in the United States that had silos. Silos became more numerous in the 1900s with the development of several different structural components including tile and concrete, which gradually replaced the wooden silos. The vertical silos also served as storage for grass silage and high moisture corn. With the increased size of dairy herds, in the late 20th century, the bunker silo came into use for both

corn and grass silage. In about the same time period, methods were developed for baling round bales of high moisture hay and wrapping it in plastic to prevent spoilage. Huge plastic bags, approximately eight feet in diameter and up to 100-feet-long, are also being used for the storage of silage.

Barns built for the housing of dairy cattle are the most prominent barns scattered throughout the State's countryside. The first barns, both the Dutch barn and the English threshing barn, were multi-purpose, providing storage for both grain and livestock. Both of these barns had gable roofs, shaped like an inverted V. The eves of the barn were not as high as later barns because all of the hay was pitched from the wagon into the mow. It was not practical to have the sides higher than could be reached by one man pitching the hay up to another man in the mow. When the horse fork was invented, which carried the hay to the top of the barn, it eliminated the need for the hay to be relayed from man to man and higher barns began to appear on farms. In the latter 1800s, when milk began to be commonly shipped to a cheese factory or creamery, dairy herd size increased. At this time gambrel roofs began to appear on barns to provide additional room for the loose hay. The gambrel roof has two pitches on each longitudinal half, with the first pitch quite steep and the second pitch much flatter, creating additional room for hay storage.

Housing for dairy cows at Merrell Farms in 2009, Wayne County

Following the adoption of the gambrel roof, there were a number of varieties of trussed barns with rafters extending from the barn floor to the peak without any obstructions from cross beams that were necessary in both the gable and gambrel barn construction. Most of these truss-designed barns were patented or were sold under a brand name. With the development of waterproof glues, laminated rafters also came into use.

The common use of silos, in the 1900s, decreased the need for some of the hay storage in the dairy barn since silage became a larger part of the cows' diet. The use of the pick-up baler on farms in the 1950s also made a big change in the size of barn that was needed. Baled hay required less storage space and barns built at that time had a lower roof because less height was needed for hay storage.

The milking parlor with free-stall housing for dairy cows brought about a huge change in barns and made the dairy barns of the previous century obsolete on most New York farms. Huge pole barns were constructed to house the dairy cows, with a much smaller building connected for milking the cows. The cow's food gradually became both corn and hay silage. At the present time, dry hay is seldom fed to milking cows but is still fed to dairy heifers. This requires only a small amount of barn storage for hay. Some of the traditional dairy barns are still used for dairy heifers and occasionally still house a herd of milking cows, but the number is declining. A rather new type of barn, popular for dairy heifers, is the "greenhouse" barn, which is a fabric-covered frame providing excellent light and air movement.

Buildings for the storage of corn have also undergone substantial changes. The small corncrib, which held whole ears of corn, was replaced by larger corncribs. With the use of the corn combine these large corncribs were replaced by steel bins, which hold many thousands of bushels. Wheat, oats and soybeans are also stored in large, metal, round bins.

With the advent of electricity to provide light during all seasons of the year, and the development of nutritionally complete feed rations, poultry production moved into large buildings. Only an occasional farm, specializing in egg production in the early 1900s, used these larger buildings but as flock size increased more of them became common. Buildings used for egg production today are very large pole

barns that are totally mechanized, with conveyers for both feeding the birds and transporting their eggs from the barn to the egg processing room. The manure from the chickens falls to the lower part of the barn where it may be stored for several months or removed daily by a conveyer.

Pole barns became popular on farms in the last half of the 1900s. They are less costly to build and are multipurpose. They are used for machinery storage, as well as for livestock housing and crop storage. Most of the barns built to house dairy cattle in the past 40 years have been large pole barns, some more than an acre in area. The manure from the cattle is automatically removed, and on many farms is transported to an aerobic digester for the production of methane gas, or to a huge tank or a specially designed lagoon for the manure's storage until it is applied to the land.

There are other barns constructed for special purposes. Potato storage barns usually have gable roofs and are insulated and ventilated to store potatoes under ideal conditions. There are buildings for the storage of onions and refrigerated buildings with controlled atmosphere storage for apples.

Greenhouses in proximity to large cities were common for over a century. With recent increased specialization in vegetable, nursery and floriculture production, greenhouses have become more common. Initially greenhouses were constructed with glass in wooden or metal frames but in recent years many are covered with plastic, which are more economical.

In 1997, the New York State Barn Coalition was formed to promote the preservation of some of our barns. There were grants made by the State government providing financial aid for barns worthy of restoration. A number of barns were saved with this program but currently it is no longer funded by the State.

FENCES
During the early years of New York State, domestic animals ran loose and crops were fenced to keep the animals out. An advantage of this practice, beyond need to protect the crops, was that cattle enjoyed the young shoots growing from tree stumps. This killed the trees and allowed the stumps to be

Stump fence in Steuben County in 1916. Stumps laid in a row, with an occasional rail, made a durable and effective fence. Courtesy of New York State Archives.

removed years earlier than if the shoots had been allowed to grow.

Settlers needed to make fences from available stone or wood, as barbed wire did not come into use until the last half of the 19th century. Building fences of stones took a great deal of time, so most stone fences were not constructed until deeper plowing exposed more stones and the forests had been removed. Stone fences were common in limestone sections of the State and areas where there was an abundance of stones. Since stones needed to be moved from the fields it was logical to use stones of appropriate size for fences. Some of the stone fences were works of art and others were huge. North of Oneida Lake there was a long stone fence wide enough on the top to drive a horse and wagon. Most of these stone fences have now disappeared, having served their purpose, although some have been reused as the walls for or in the landscaping of a suburban home. Sometimes we find an old stone fence winding through the forests where originally crops were grown.

The first wooden fences were called rail fences. Rail fences were made from small logs that were split with axe, beetle and wedges of wood, called gluts. Some rail fences were zigzag, eliminating the need to dig holes for posts in the ground. Others used X bars to support the rails, which also did not require posts in the ground. When a post in the ground was used, the post was mortised to accommodate either three or four rails.

The rail fences were not as animal-proof as our woven wire, barbed wire or electric fences of today. Very often a triangular or "p" shaped wooden poke, significantly larger than the animal's head, was placed around its neck to prevent the animal from slipping through the fence.

Pine trees thrived on the sand and gravel soils found in many parts of the State. The soils where pines thrive are normally more acidic (lower ph). After the trees had been cut and the farmer tilled the land around them a few years, he dug around the stumps and pulled them out with his oxen. They were then dragged into rows around the fields he wished to fence. The stumps, pulled closely together, made a substantial fence that animals could not pass through and which lasted for many years. Some of these fences were still in use in the 1920s. Occasionally a farmer would build or buy a stump puller. Henry Dabold of Van Buren, in Onondaga County, was one of the first to use a stumping machine. By using a sweep powered by horses or oxen, the puller, which set over the stump, exerted a force many times greater than pulling directly on the stump.

In the late 1800s, as our country became more industrialized, barbed wire came into common use and the rail, stone and stump fences gradually disappeared. Special fences of wire have been made to confine poultry, livestock and even deer. During the last 50 years, the electric fence has become common because of its low cost and the convenience of easily moving it from one location to another when rotating cattle pasture.

DRAINAGE

The drainage of excess water from farmland was not a serious concern when New York farms were first settled. A home site was selected on a knoll or where the ground was higher than the surrounding land to avoid potential flooding in wet periods. The land that was initially cleared was in proximity to the home site and where the trees were the largest, which was indicative of rich soil. Land that was excessively wet or hilly was left to become permanent woodlot, pasture or delegated to be cleared at a later time. In the late 1700s and early 1800s, there was plenty of land available and it was cheap. A settler was not inclined to spend extra money improving his land with so much land available at a low price.

As our New York lands became settled and farm-land appreciated in value, more attention was paid to draining both wet spots and entire fields where excessive water limited yields at certain times. Little is mentioned in early 1800 farm literature concerning drainage, but by the 1840s it became a hot topic. There were occasional farmers draining wet spots in their fields at an earlier date, but it was uncommon and the methods were often primitive. An early example is Anson Sweet of Pompey, in Onondaga County, who in 1818 used underground draining on his farm.

There were a considerable number of variations in underground draining. Various types of stone ditches were laid on farms where there was a ready access to stones. Often it was considered to be a good method of getting rid of stones. A trench was dug, varying from 2 ½ to 4 feet deep, where water could easily flow to an outlet. A channel was formed, for water to flow in the ditch bottom, by laying a row of stones on each side of the bottom of the ditch, leaving a space in-between the two rows and placing flat stones across the top. Normally, some straw or fine brush was laid on the top of the stones before filling the trench with dirt, preventing dirt from finding its way into the channel before the dirt had time to settle. The author has repaired underground drains, constructed by the latter method over 100 years earlier, numerous times. Another method of making stone drains was to place a row on only one side of the ditch bottom and then place flat stones, leaning against them at a 45 degree angle, forming the left side of an "N". A third method of making a stone ditch was to throw cobblestones in the ditch to a depth of at least one foot, cover with straw and fill with dirt. This method took less time but was not as effective as the previous two.

Sometimes the ditch was dug the usual width and then the center of the bottom was extended 5 inches deeper making an opening 5 inches square. Rye straw, which is longer than the straw of other grains, was twisted by hand into what resembled a five-inch rope, placed into the opening and covered with dirt. The water found its way through the strands of straw but it only drained small quantities of water rather slowly. Boards were also often used, with hemlock the most desirable because of its resistance to rot. Two boards were fastened together to form an inverted V, which was sometimes placed on a board at the bottom and other times a third board wasn't used depending upon soil conditions.

In 1821, John Johnston of Scotland, came to the town of Fayette in Seneca County and purchased a 112-acre farm on the shore of Seneca Lake. He put in the first clay tile drain in America in 1835, buying the tiles in Scotland and shipping them to New York. Many farmers scoffed at spending money to improve land when so much land was available at a low price. By 1851, he had laid 12 miles of tile on his farm.

Because of significantly increased yields from the fields where Johnston put in his underground drains, other farmers began to copy his methods. Tile factories were later opened in nearby Waterloo, to meet this demand for tile. Before a tile-making machine was available in New York, some of the first tiles were called shin-tiles, formed by a man placing the clay on his shin and then removing it for baking in an oven. This tile resembled a horseshoe and later thousands of tiles were made by machine in this horseshoe shape. Forming the clay around a pole and removing the pole from the clay before firing produced a tile with a round hole in the center. In 1848, John Delafield, a neighbor of Johnston's, imported a tile machine from England that decreased the cost of the tile and improved the quality. Delafield was a strong believer in tile and promoted their use as much as Johnston.[1]

John Johnston is considered the "father of tile drainage", in the United States. Today, there is the Mike Weaver Drain Tile Museum at the former home of Johnston, close to Rose Hill Mansion near Geneva.

An article in the 1847 *Transactions of the New York State Agricultural Society,* describes the draining of a 10-acre tract of land at Hells Gate Neck, adjoining the village of Astoria on Long Island. It said they turned a quagmire into a productive piece of property. They used the method earlier described of two rows of stones with a layer of flat stones on top. Articles in the 1859 book, *Transactions of the New York State Agricultural Society,* tell of a Geneva farmer putting in 50 miles of tile on 200 acres for a total

1   Hendricks, Ulysses, Prentiss *A History of Agriculture in the State of New York* p. 348-350

cost of $5,000 or two cents a foot. Another article in the same book, pertaining to another farmer, indicates the going cost was 31-34 cents a rod, which is two cents a foot. These costs were low and with the advent of the Civil War, the cost of tiling increased.

For many years, farmers debated which method was best: stone or clay tile or open ditches. For farmers on heavy clay soil, the only successful method was often shallow, open ditches for removing excessive water from the fields. Open ditches are also the common method for draining fields of muck.

Clay tile became the standard for under-draining until the availability of perforated plastic tile in the middle of the 1900s. Plastic came in long rolls and was much more convenient to install. With the invention of the laser, a large tractor, with a tile-laying attachment, is able to lay thousands of feet of tile, placed at an exact grade, in one day. Draining fields has come a long way from the time they were dug by hand after oxen had pulled a walking plow opening the upper part of the ditch.

TABLE 33 – Number of New York Farms and Acres with Irrigation
US Census

Year	Number of Farms	Acres Irrigated
1940	567	5,948
1950	888	19,248
1978	1,715	56,106
1997	2,501	69,197
2007	3,036	68,010

New York has a great variety of soils and terrain. There is no question that draining has improved and made possible the use of thousands of acres of fine New York farmland. Some areas of the State with well-drained soil have little need for underground drainage, but in other areas the land would be almost worthless without drainage.

New York is blessed with a climate that normally provides sufficient rainfall to produce a reasonably good crop. In almost any year there are portions of the State that receive too little rain at the right time to produce a bountiful crop, while there are other

areas that receive too much. The cost of irrigation is generally greater than the additional return that could be gained by irrigation of hay and grain crops. Irrigation often pays large dividends for those crops with a high market value like market garden, nursery and floriculture crops. **Table 33** shows the number of farms irrigating land and the acreage of crops irrigated since tabulation of irrigation information by the U.S. Census began in 1940.

Equipment for irrigation was limited prior to 1940, but since that time the number of farms with irrigation has increased six-fold and the number of irrigated acres ten-fold. Water available for irrigation comes from a variety of sources including lakes, rivers, deep wells, ponds and municipalities. New York is fortunate to have a large supply of fresh water available for irrigation over wide areas of the State. Suffolk County has 377 farms with 13,614 irrigated acres and Genesee County has 40 farms with 7,832 irrigated acres. Other leading counties with irrigated land are Columbia, Orleans, Ulster, Orange, Dutchess and Erie. Genesee and Orleans farms tend to be larger because they produce vegetables for processing plants, whereas the farms in the other counties are in close proximity to metropolitan areas and are of smaller size producing food, floriculture and nursery items that are sold directly to the retail customer or to dealers selling to the retail customer.

FERTILIZATION
The introduction of the 1860 U.S. Census states," land is abundant and cheap; labor is scarce and dear". This statement sums up the prevailing attitude of the majority of farmers at that period of time. Although the bulk of New York's farmland had been cleared by 1860, much of the soil's original fertility remained and farmers were satisfied with their yields without purchasing fertilizer. A large portion of our farmland remained in permanent pasture for many years and did not have the nutrient drain that might have occurred with continual tillage. In addition, most of our farms were general farms, with livestock providing animal waste for fertilization and with a variety of crops grown, there was considerable crop rotation.

Not all of our farmers were oblivious to the need of returning nutrients to the soil removed by crops. The majority of our farmers had roots in Europe and some had recently emigrated from Europe and were familiar with the agricultural practices there. In Europe the land was neither abundant nor cheap and soil fertility maintenance practices had been instigated earlier. Some New York farmers were using field trials to determine the best methods of maintaining soil fertility. One example was Joseph Harris of the Rochester area who, in the 1850s, ran experiments measuring the effect of fertilizers upon corn yields. He added various amounts of wood ashes, guano, super phosphate and other potential fertilizers to measured plots of ground and compared the yields to a control plot.

In articles in the 1819 and 1820 *Plough Boy* magazines, printed in Albany, there are several articles that go into great depth regarding such items as night soil, animal manure, bones, limestone, gypsum, crop rotation and green manures. These articles all come under the heading of "manures", which would indicate that at that period of time "manures" covered all the currently available methods of maintaining soil fertility.

Until the late 1800s, New York had no Agriculture Experiment Station or agricultural college to conduct experimentation and relay new information to the farmer. Farmers learned from their neighbors, their own experiments, at fairs and by reading farm magazines. Often the information received was not scientifically reliable, and normally there was considerable resistance to suggested changes in farm practices. Because of these factors, most farmers did not purchase any appreciable amount of fertilizer until into the 20th century. In 1875, the total cost of fertilizer purchased by farmers was $1¾ million with two-thirds of that amount used on Long Island. Long Island, which was producing vegetables for New York City, needed more fertilizer because its soils had been producing crops for 200 years and nutrients leached from its sandy soils more rapidly than from the heavier upstate soils. In 1880, New York farmers purchased less than $3 million of fertilizer and by 1900 purchases were slightly less than $4½ million, making the average cost of fertilizer per farm $20. It is very likely that a minority of farmers were using commercial fertilizer at that time and those that were, were the larger, more progressive farms.

By 1920, fertilizer purchases passed $15 million but 20 years later, because of the Depression and low prices for farm products, fertilizer purchases decreased to less than $9 million. In times of low prices for their crops, farmers, often of necessity, cut back on their purchases of fertilizer. By 1969, fertilizer expense had increased to $33 million and only nine years later it more than doubled to $87 million. The increase in the cost of fuel coupled with the economic inflation of the 70s, caused higher farm prices and an increased use of commercial fertilizer. The cost of fertilizer purchases remained fairly level for a decade, but in 2007 the cost of fertilizer used by our New York farmers passed $172 million. Part of the increased cost was inflation and part was because farmers had, in the previous few years, received improved prices for their products.

Organic fertilizer, from animal waste, cover crops and crop residue, was the major fertilizer used on our farms into the 1900s. Organic fertilizer is still critically important and is abundantly produced and utilized on dairy farms. Crop yields have increased dramatically during the past century and like animals, crops have to be fed well to obtain the greatest yields. A large crop is necessary to realize a profit and this requires farmers to purchase additional commercial fertilizer. The use of commercial fertilizer is likely to increase in the future as the population of the world increases.

CHEMICALS

Insects and plant diseases have plagued farmers since plants were first planted by man. In the 1820 *Plough Boy* there are numerous articles suggesting how insect infestations and plant diseases might be combated. Wheat seed was soaked in a variety of solutions of noxious ingredients to control smut. Tobacco dust was placed in the granary to kill weevils. Some of the attempted remedies helped and many more did not, but farmers continued to experiment hoping to find solutions.

Simple, natural chemicals have been used to help protect plants and animals since the beginning of agriculture. Man naturally wants to protect what is his and utilizes what is available to him at the time. Chemicals that were effective were very limited before the advent of research institutions and chemical companies. Without the advancements in research and its application, we would be unable to feed today's world population.

There was no census information regarding chemical use on farms until sample information was obtained in 2002. The 2007 U.S. Census provides information from all of the farms contacted showing that New York farmers spent over $103 million that year on chemicals for the control of insects, weeds, grass and brush, nematodes, diseases in crops and orchards and chemicals to control growth, thin fruit, and ripen or defoliate. 12,000 farms used chemicals on over 2 million acres.

The New York State Integrated Pest Management Program (IPM) was initiated in 1985. The New York State Department of Agriculture and Markets and The Department of Environmental Conservation, along with regional and federal grants, funded a program to minimize the use of agricultural chemicals. It was designed to operate within four agricultural areas: vegetables, livestock/field crops, fruits and ornamentals. Insects are monitored to determine if spraying is needed and if needed, the ideal time and the minimum chemical required to be applied. Insect traps are used to rid the crop of insects without sprays and insects that are natural destroyers of pests are employed. Cover crops are developed that will control weeds, and a variety of other devices are used to control disease and pests without the use of chemicals or with minimal chemical usage.

In 1994 the scope of the program was extended to all areas of pest management not covered by the Agricultural IPM. This includes schools, golf courses, parks and any other place where chemical use might be eliminated or reduced. The IPM program has eliminated large quantities of chemicals from our environment. It has also helped farmers avoid spending millions of dollars for chemicals.

Chapter XIII
Agricultural Organizations

There have been hundreds of agricultural organizations in New York during its over 200 year history. Farmers, as needed, have formed groups to solve mutual problems, to learn from each other and to enhance the profitability of their farms. There have been organizations for almost every farm product grown in New York. In the earlier years of the State, an organization might have members from just one or two towns in a county, but with improved travel, members came from gradually increasing distances, and an organization might evolve to state or even national membership. Only a few organizations will be described in this chapter, with the realization that some that are or have been of significant importance will be omitted. In addition to these, there are farm organizations mentioned in the chapters pertaining to the various farm commodities and the 2009 members of the Empire State Council of Agricultural Organizations, Inc. are listed in the Appendix.

THE NEW YORK STATE AGRICULTURAL SOCIETY

For generations, agricultural practices were passed from father to son. Observations of a neighbor's success or failure, as well as their own, were the means to improvement. Farmers' Almanacs had been printed in the late 18th century but had minimal information regarding methods of farming. Agricultural societies came into existence to share information about improved farming methods. The first agricultural society in the United States was formed in New York on February 2, 1791.

It was named "The Society Instituted in the State of New York for the Promotion of Agriculture, Arts and Manufactures" and in 1792 it printed reports of its transactions with additional reports in 1794, 1798, 1799 and 1801. Copies of these reports are very rare but were considered to be excellent in the agricultural information presented. A few of the topics discussed in the 13 meetings they held were: manuring land with seaweed, improving poor land by sowing red clover seed, cast iron plowshares and diseases of fruit trees.[1]

The charter for this first society expired by limitation in 1804, but its charter was revised and modified for it to become, "The Society for the Promotion of Useful Arts". The charter was changed because it was decided that the business of the Society could not be well-conducted at annual meetings, so instead a standing council was provided to carry on the business. The constitution stated that its chief aim was to improve agriculture and it published four reports dealing with agricultural issues. The list of members of this organization, as well as that of the first, included many prominent New Yorkers including John Jay, author of New York's constitution and second governor of New York, and George Clinton, first governor of New York and Vice President of the United States. The Society was succeeded in 1819, partially by a Board of Agriculture established by an act of the Legislature and partially, in 1829, by merging into the Albany Institute.[2]

---

1   Hedrick, Ulysses Prentiss *A History of Agriculture in the State of New York* p. 113, 318
2   Ibid. p. 118

Elkanah Watson, promoter of Agricultural Fairs and an agricultural college, circa 1825.

The State created the Board of Agriculture with an annual appropriation of $10,000 to be distributed among county agricultural societies for holding county fairs. The various county agricultural societies were empowered to elect officers who were "practical farmers" and these officers were to "annually regulate and award premiums on articles and productions as they may deem best calculated to promote the agricultural and manufacturing interests of this State". There was also a provision that each person receiving a premium for an agricultural product should make a full description of the methods he had used in producing it. The ex-presidents were to meet annually in Albany to examine these reports and publish the ones that seemed worthy. The Board of Agriculture continued until 1826, when it expired by statue limitation.[3]

There were six counties that established agricultural societies in 1819. These six agricultural societies,

along with numerous others founded in succeeding years, played a significant role in promoting improved methods of agriculture. The major means used to transmit valuable information to the farmer was fairs. Fairs gave farmers the opportunity to see the best animals and products from farms, new labor saving tools and improved production methods. Newly-invented machines were displayed and competed against each other in field trials. The opportunity for farmers from different areas to get together to see new ideas and discuss opportunities and challenges was also invaluable.

Elkanah Watson, a promoter of public enterprises, could be considered the founder of agricultural societies and fairs. He was active in both Massachusetts and New York. In 1811, he organized a successful fair for Berkshire County in the village of Pittsfield. He returned to New York in 1816 and was asked to organize a fair in Cooperstown for Otsego County that was also successful. There was another fair in Cooperstown the next year and also one in Jefferson County, both organized by Watson. Word of his successful fairs rapidly spread and he was asked to help put on fairs in Oneida, Schoharie, Montgomery, Rensselaer and New York counties. He gave speeches promoting agriculture at each of these fairs, and undoubtedly his efforts were a catalyst in the New York Legislature's decision to form the Board of Agriculture and provide the $10,000 annually for county fairs. Watson was also an early promoter of an agricultural experiment station and an agricultural college, but it was about 50 years before these came into existence.[4]

Members of the former Society for the Promotion of Useful Arts and those persons who had been associated with the Board of Agriculture felt the need for a new agricultural society, and in 1832 the New York State Agricultural Society was founded. Its mission was "to improve the condition of agriculture, horticulture and household arts". Le Ray de Chaumont was elected president and Jesse Buel, secretary. In 1834, Jesse Buel established a newspaper called *The Cultivator* and the Agricultural Society named *The Cultivator* as the official organ of the Society. *The Cultivator* became one of the most influential agricultural papers of its time. When Jesse died in

3   Hedrick, Ulysses Prentiss *A History of Agriculture in the State of New York* p. 122

4   Ibid. p. 122-132

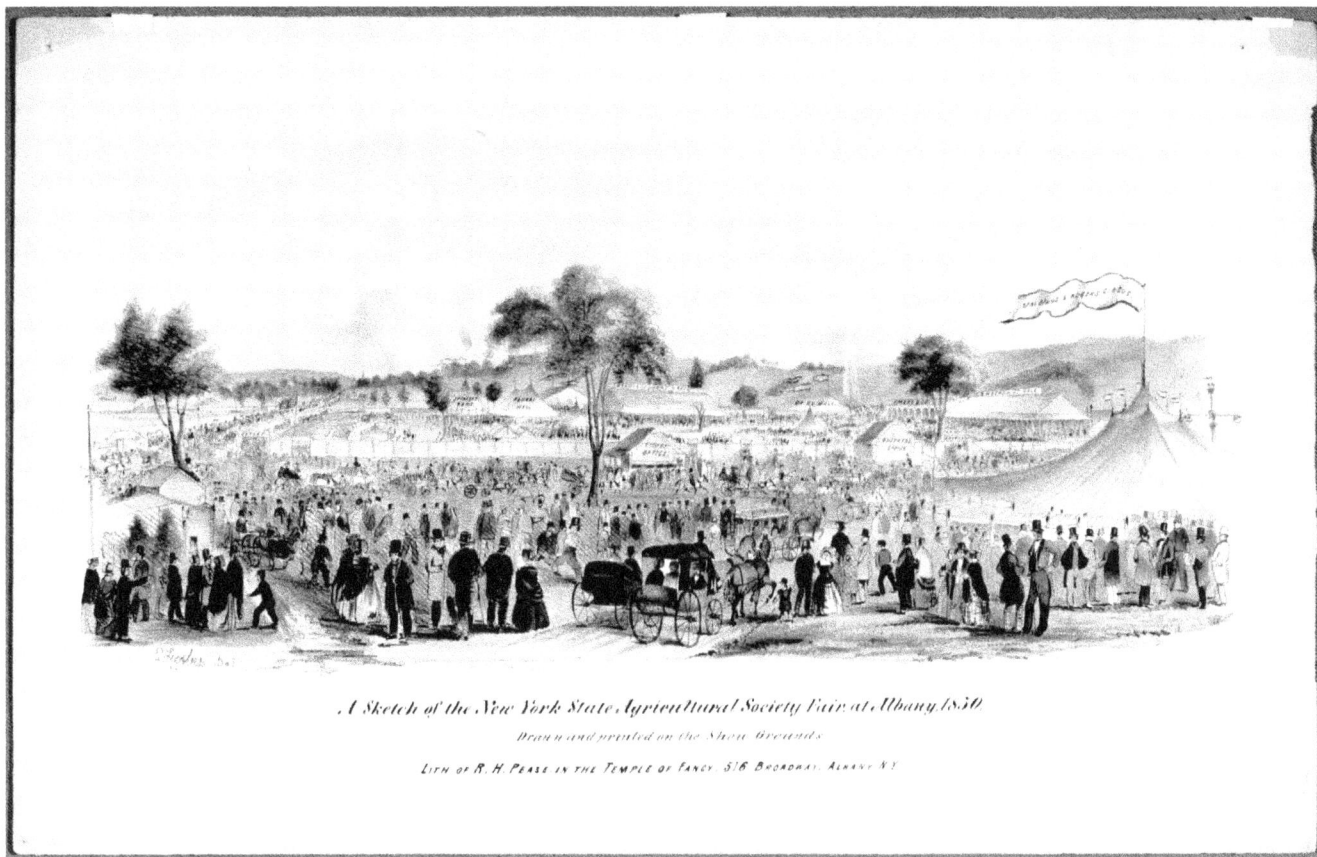

A sketch of the New York State Agricultural Fair at Albany, 1850. Courtesy of Albany Institute of History & Art u2005.13.

1839, it was united with the *The Genesee Farmer*, and in 1866 was merged into *Country Gentlemen*.[5]

The written word had been used to disseminate agricultural information for centuries, but was not often of value to the common farmer because of cost, difficulty in transmittal and the limited availability of education in rural areas. George Washington, on his Mount Vernon farm, did a great deal of experimentation and regularly wrote letters to English farm authors. These letters were published in the early 1800s and dealt with such topics as farm machinery, livestock, crop rotation and fertilizers. Although his work was done in Virginia, much of it applied to New York and other states as well. Agricultural journalism in New York began in 1819 when Solomon Southwick launched the newspaper *The Plough Boy* in Albany, publishing it for 20 years. In 1831, Luthur Tucker published *The Genesee Farmer*, and in 1853 *Country Gentlemen*. These papers were popular and contained significant agricultural information,

but were only a few of dozens of papers and magazines, relating to agriculture, that were published in the State during the 19th century.[6]

At the 1841 annual meeting of the New York State Agricultural Society, a new constitution was adopted to position the society to have a more positive effect on improving the condition of agriculture in the State. Because of this change, the New York Legislature appropriated $8,000 per year for five years "for the promotion of agriculture and household arts in the State". All of the money was to be distributed to the counties of the State with the exception of $700 that was given to the Society for its own use.[7]

Fairs were a major educational institution for New York farmers in the 19th century, prior to the advent of experiment stations and agricultural schools. It is appropriate to include some information about the New York State Fair since the New York State Agricultural Society operated the State Fair from its beginning in 1841, until 1900 when the State took

---

5   Hedrick, Ulysses Prentiss *A History of Agriculture in the State of New York* p. 119-120

6   Ibid. p. 321-326

7   Ibid. p. 121-122

New York State Fairgrounds circa 1970, prior to fire and subsequent removal of the Machinery Building. Courtesy of New York State Archives from New York State Ag & Mkts.

over its operation. The first New York State Fair was held in the village of Syracuse on September 29 and 30, 1841 on North Salina Street near Ash Street. There were many premiums offered for agricultural and household-made items. The day before the Fair opened, a whole train arrived from Albany with 25 cars of choice livestock for display. There were 15,000 people in attendance and a plowing contest was held in nearby Onondaga Valley. A farmer's dinner was held with 1,200 people in attendance. In an exhibition of farming equipment there were threshing machines, straw cutters, fanning mills, plows, harrows, cultivators, drills, cradles, scythes, pitch forks and horse rakes. The Fair was held in Albany the next year and was extended to three days. The third fair was held in Rochester with 18 entrants in the plowing competition. Three of the numerous notables present were ex-president

Martin Van Buren, Daniel Webster and William Seward. In the following years, until Syracuse was selected as a permanent site in 1890, the now annual State Fair was held at Elmira, Utica, Buffalo, Saratoga, Watertown, New York City, Poughkeepsie and Auburn, in addition to the three cities mentioned previously.

A site of 100 acres, pastureland of the Smith and Powell Stock Farm and Nursery, was selected west of the city of Syracuse at a cost of $30,000 and offered to the New York State Agricultural Society. The Society accepted the offer and the first Fair held on this site was in 1890. During the next decade, the Fair was not an outstanding success and in 1900 the Society and the State mutually agreed that the State would accept the property and operate the Fair in the future. Numerous changes were made over the succeeding years including approval by the

State legislature, in 1900, of $10,000 for a mile track and in 1923, funds for the Coliseum. The Indian Village was added in 1928 and other improvements followed. A move to add more amusement features along with the educational aspects of the Fair increased the attendance and its success.

The New York State Fairgrounds were used by the US Army during both World Wars. In January 1944, a dike broke at nearby Solvay Process and much of the fairgrounds was flooded with waste product. It did a great deal of damage and consideration was given to moving the Fair to another location. It was decided to continue the Fair at its existing location and in 1948 there was a partial fair on the site. In 1960, Fair attendance passed the 500,000 mark and in 1962 the official name of the Fair became the New York State Exposition. The New York State Exposition remains on the same site today. In recent years, close to one million people have visited the Exposition each year.

During the middle of the 19th century, the Ag Society had established a museum and library near where they met in the Capitol in Albany. The museum items became part of the larger State Museum and in the 1920s the Ag Society encouraged the State to build an agricultural museum on the New York State Fairgrounds. In 1928, the State provided funds to do so, with the Museum named for Daniel Parrish Witter, a prominent farmer and member of the state legislature. Later, a carriage museum was constructed nearby, named after Jared van Wagenen, Jr., a prominent agriculturalist. In 1996, the Board of Directors of the Agricultural Society formed a committee to work with the Fair administration to improve the Museum and to develop high quality educational exhibits. This committee formed a non-profit corporation, the Friends of the Daniel Parrish Witter Agricultural Museum. This group is currently working with Fair Director, Dan O'Hara, Department of Agriculture and Markets (2009) Commissioner, Patrick Hooker and State legislative leaders to bring the Museum up to 21st century standards as an educational facility, not only for the farm families of the State, but also to educate city and suburban residents in both the rich history of New York agriculture and its continued importance today.

Each year hundreds of cattle, horses, sheep, swine and poultry are exhibited at the Fair. Farm youth are active participants though 4-H and the Future Farmers of America (FFA) programs. In 2007, the Fair increased emphasis on agriculture and its importance to New York's economy, in the tradition of the New York State Agricultural Society and the State Fairs during the 19th century. The Fair is taking steps to improve the visibility of agriculture. A Pride of New York display showcasing some of New York's agricultural products, greets Fair visitors as they enter the grounds. Recognizing the importance of youth to New York's agriculture, improvements have been made to the 4-H, Cooperative Extension and youth dormitory areas.

The New York State Fairgrounds are bustling with activity, much of it related to agriculture, all year round. There are about 30 horse shows annually, ranging in size from 25 to 400 horses. In April of each year there is the Dairy Carousel, which is the second largest cattle show in the country. Each February there is a huge Farm Show with thousands of people coming to see the latest developments in farm equipment and supplies. In addition,

The Daniel Parrish Witter Agricultural Museum at the New York State Fairgrounds.

throughout the year there are special shows for goats, alpacas, rabbits and a myriad of other activities. All of this is in addition to the approximately 10,000 animals that are exhibited during the 12 days of the Fair. Today the Fair is not the educational mechanism for teaching farmers that it was in the 19th century, but more a means of bringing an understanding of New York's agriculture to the consumer. This is appropriate as today there are numerous other ways of providing useful information to the farmer.

Ezra Cornell, founder of Cornell University, recognized the importance of fairs and the New York State Agricultural Society in his 1863 address, as the Ag Society's retiring president. *"Much of this improvement* (mechanization and improved efficiency in agriculture) *and many of these new inventions may be traced to suggestions or encouragements held out by this Society, or to ideas or thoughts which were quickened into active inquiry and directed to inventive channels by visits to our annual Fairs, or occasional implement trials."*

An additional important educational service of the New York State Agricultural Society was the printing of the T*ransactions of the New York State Agricultural Society*, annually from 1841 to 1898. These books included reports of the annual fair, activities of the county agricultural societies, illustrations of innovative agricultural ideas and research reports from all corners of the world.

Examples of the interesting pieces of information in these annual books comes from the *1850 Transactions of the New York State Agricultural Society* regarding the 10th annual fair held in Albany. Attendees at the Fair came from 22 states and 4 Canadian provinces. In the plow trials, 22 plows were tested on stiff sod, 21 plows on sandy sod and there were 3 side-hill plows. The cost of the plows varied from $7.50 to $28 and weighed from 100 to 175 pounds. Among the power tools shown were a dog-powered treadmill for a butter-churn, 2 and 4 horsepower sweeps and a ½ horsepower steam engine selling for $75. During the previous year, the Agricultural Society had made contributions to the State Museum of plows, seeds and books. At that time farm labor cost 50 cents a day and a hired girl earned $1 a week. There was also an in-depth, 200 page section of the book regarding agriculture in Seneca County. There was a wealth of other information included,

as this was a period when farmers did their own research and shared it with each other.

Few people, other than farmers, recognize the significant role of the New York State Agricultural Society in the development of New York State agriculture. It has largely been a behind-the-scenes player, making things happen for the betterment of agriculture. Today, it still has an educational annual meeting attended by hundreds of interested agriculturalists and a Board of Directors that meets twice a year. Some of its other accomplishments, in addition to initiating the New York State Fair and operating it for nearly 60 years, are: encouraging the appropriation of funds, by the legislature, leading to the establishment of the many county fairs; support for the establishment of Cornell University and its agricultural college; supporting the Geneva experiment station and the experiment station at Cornell; involvement in the establishment of the New York State Dept. of Agriculture and Markets and in the construction of the Witter Museum on the New York State Fairgrounds.

The work of the New York State Agricultural Society has continued through the years with a Century Farm program honoring families that have operated the same farm for more than 100 years. Since this program originated in 1937, more than 340 Century Farm families have been honored. A number of farms have been owned and operated by the same family for more than 200 years. In 1956, the Ag Society instituted the Distinguished Service Citation, yearly honoring a person who has provided distinguished service to agriculture. Beginning in 1975, a Journalism award program was instigated honoring outstanding journalists for their coverage of agricultural news. In 1984, the Ag Society initiated another significant program with the establishment of the Empire State Food and Agriculture Leadership Institute (LEAD New York). It was developed in cooperation with Cornell University and is a two-year leadership-training program for 30 men and women working in agricultural related fields. This program has been a tremendous success. Its 12th class graduated in 2009. These and other positive programs are the heritage of the "Ag Society", an organization that continues to work for the betterment of New York's agriculture.

Conference of Agricultural Organizations meeting with Governor Mario Cuomo, circa 1990. Courtesy of New York State Department of Agriculture and Markets.

EMPIRE STATE COUNCIL OF AGRICULTURAL ORGANIZATIONS, INC. (CAO)

The CAO was formed in 1954 to enable its members to develop a united platform on fundamental economic, social, legislative/regulatory and other state and national issues affecting agriculture in New York State. CAO is built on collective strength and policy statements that are unanimously adopted by its entire membership, representing numerous facets of New York's agricultural industry. A list of the 2009 members of the CAO is provided in the Appendix.

The name of the original organization was the New York State Conference Board of Farm Organizations. There were eight members: New York State Grange, Dairymens' League, New York State Federation of Home Bureaus, New York State Horticultural Society, New York State Vegetable Growers Association, Cooperative Grange League Federation Exchange, Inc., New York State Poultry Council and the New York State Farm Bureau Federation.

The formation of the CAO was an important step in the progress of New York agriculture as it provided a unified voice for agriculture with the strength that exists in large numbers. Previously, organizations spoke only for themselves, and while they may have had similar goals, their presentations varied, often without the success obtained by unification.

In the early days of the organization, recommendations of desired State legislation or funding requests were made directly to the Governor, appropriate legislative representatives and the Department of Agriculture and Markets. In recent years, direct presentation of the CAO's agenda continues to legislative leaders and Ag & Mkts, and indirectly to the Governor, through his aides.

Almost every important piece of legislation affecting New York's agriculture during the past 55 years has been proposed by the CAO. It has also been equally important in opposing actions that would be detrimental to New York's agriculture. The formation of the New York State Farm Viability Institute, established a few years ago, occurred through the cooperative efforts of several CAO members. As New York's agriculture changes, the membership of the CAO changes, but its importance to New York agriculture continues.

GRANGE

The National Grange of the Patrons of Husbandry was formed in 1867 in an effort to aid the rural south in recovery from the disastrous effects of the Civil War. Within a decade the Grange, as it became known, had over 750,000 members nationally with over 11,000 in New York. Initially, membership was largely from the southern states, although there was a subordinate Grange formed in Fredonia,

Grange Hall at Henrietta in Monroe County in 1913. Courtesy of New York State Archives.

New York in 1868. Gradually the Grange movement came to the north and the New York State Grange was formed in 1873.[8] By 1915, there were more than 1,000 local granges in New York with a total of over 100,000 members.

The rapid growth of the Grange was largely due to the fact that farm people believed they were unjustly suffering from excessive railroad freight charges on their produce. Farmers also believed the prices they had to pay for the supplies they purchased were excessive and the opportunity for cooperative purchasing power was appealing.[9]

The Grange has had an illustrious record of service to rural residents at the local, state and national levels. Many of the local Granges established Junior Granges to benefit the children of their members. These encouraged honor and patriotism, and fostered the development of self-confidence, responsibility and respect. The young members had their own special projects and earned merit badges for their work.

Discussions on critical topics of the day were held in the local Granges with recommendations on state and national issues sent on to the county, state or national levels where policy is formed and presented to government policymakers for their consideration. A few of the many issues promoted by the Grange have been rural electrification, improved roads, legislation to control the railroads, the elevation of the USDA to Cabinet position, establishment

of agricultural colleges and agricultural experiment stations, teaching agriculture in public schools and Rural Free Delivery. Community service has also been an important function of the Grange.

A major service provided to New York agriculture by the Grange was its strong philosophy, from the very beginning, that the cooperative effort of many farmers would bring greater success for each farmer. This philosophy encouraged the growth of hundreds of cooperatives, both in New York and beyond, that were beneficial to all in agriculture. In the following paragraphs are some examples of the Grange's involvement in the cooperative movement. There is also information in the section on the marketing of dairy products in the next chapter regarding the Grange's involvement in the formation of The Dairymen's League.

The first Cooperative of Trade organized by the Grange was established in Monroe County in 1874. Initially, it served only members of local Granges in Monroe County but soon Grange members in other counties were participating in savings through the large purchases it made with small fees charged to the members. The Cooperative sold Grange members a wide variety of supplies including farm machinery, lumber, feed, seeds, fertilizer and coal. When it was formed there was strong opposition from established businesses selling supplies directly to the farmer, because of their fear of losing business to the Cooperative. By 1881, sales to farmers exceeded a million dollars. All of the members were required to pay cash for their purchases.[10]

The Grange Exchange, a capital stock company with stock of $100,000, was formed in 1919. It, like its earlier cousin, received opposition to its formation. The New York Feed Dealers Association managed to cut off the supply of feed to the new company, but The Grange Exchange eventually made arrangements with some grain mills to purchase feed. The Grange Exchange also sold fertilizer and seeds with sales of 13,000 bushels of seed corn in its first year of operation.[11]

In June, 1920, the New York State Grange, the Dairymen's League and the State Farm Bureau Federation combined to form the Grange League Federation (GLF), which succeeded the Grange

8   Alexander, L. Ray *History of the New York State Grange 1873-1933* p. 19-20

9   Ibid. p. 20

10   Ibid. p. 121-122

11   Alexander, L. Ray *History of the New York State Grange 1873-1933* p. 124-125

Exchange. The GLF, which later became known as Agway, was successful from its very beginning with sales of over $25 million by 1928. H.E. Babcock served as its general manager until 1932, when he became president of the company. In 1929, it boasted of opening the largest and most up-to-date feed mill in the country at Buffalo. It also operated a chain of over 100 retail stores in the New York milk shed. Within a few years it developed a large seed business, fertilizer factories, a large feed mill in Albany and even a paint factory.[12]

The Grange pioneered the cooperative movement in insurance as early as 1874 with an effort to organize a mutual life insurance company in Elmira, New York called the Patrons Aid Society. It had 400 members within two years, but only survived about a dozen years. About two years later, the Grange organized Grange mutual fire insurance companies and by 1878, there were 25 of these companies doing business. By 1882, there were 60 mutual insurance companies doing business in the State. In 1913, the state granges of New York, PA and Ohio formed the Farmers & Traders Life Insurance Company, which grew very rapidly, providing life insurance for Grange members in 20 states.[13]

In 1973, a new State Grange Headquarters building was dedicated in Cortland, which has served as offices for other agricultural organizations as well. In 2009, there were 236 Grange Chapters with 7,000 members in New York. The Grange continues its long history as a proponent for the needs of farm families and the residents of rural communities throughout New York.

FARM BUREAU

The official beginning of the New York Farm Bureau, as a state organization, was in 1917 with the formation of the New York State Federation of County Farm Bureau Associations. The very beginnings of Farm Bureau, in New York, occurred six years earlier in Binghamton when John H. Barron was hired as county agent. During the previous year the secretary of Binghamton's Chamber of Commerce, Byers H. Gitchell, began agitation for a department in the chamber devoted to "extending to farmers the same opportunities for cooperation now enjoyed by

the business men of this city".[14] Gitchell's concern came from a report of the Country Life Commission, chaired by Liberty Hyde Bailey, and of the interest shown by James Wilson, Secretary of Agriculture, regarding abandoned farms on the hills of Southern New York. The idea was also encouraged and aided by the traffic manager of the DL&W Railroad. The Chamber appointed a committee, which decided that a "farm bureau" should be established with a farm agent in charge, and that the Chamber, the Railroad and the United States Department of Agriculture should finance the enterprise.[15]

The State College of Agriculture was unable to contribute financially but was to give advice and encouragement. Barron, the county agent, initially covered, by horse and buggy, an area within a 50-mile radius of Binghamton, which included five counties in New York and one in PA. A year later the New York State Legislature passed legislation authorizing the Broome County Board of Supervisors to make appropriations for this new farm program and the County contributed $1,000.

The Farm Bureau was having success in Broome County and by 1913 nine other counties in New York had formed farm bureaus. The College of Agriculture at Cornell, in cooperation with the United States Department of Agriculture, appointed a state leader of county agents.

In 1914, the Smith Lever Act was passed resulting in Federal-State cooperation for service to rural people and in the organization of county farm bureaus, home bureaus and 4-H club programs. By the end of 1916, there were 36 counties in New York that had farm bureaus and in 1917 the New York State Federation of County Farm Bureau Associations was formed. New York was the third state to have an organization of this kind. H.E. Babcock became the state leader of county agents and also state director of farm bureaus. The purpose of the new organization was to promote, protect and unify the work of the associations in the counties; to give them statewide representation and to speak for them on statewide issues.[16]

12   Ibid. p. 125-126
13   Arthur, Elizabeth L. *The History of the New York State Grange 1934-1960* p. 138-139

14   New York Farm Bureau, Inc. *History of New York Farm Bureau* p. 4
15   Ibid. p. 5
16   New York Farm Bureau, Inc. *History of New York Farm Bureau* p. 8-10

New York initiated the formation of the American Farm Bureau Federation by holding a meeting at Ithaca, in 1919, for representatives of 12 states, and proposing a national federation. Later that year, representatives from 31 states met in Chicago where a temporary national organization was formed and a constitution adopted. At the first annual meeting in Chicago, early the next year, the American Farm Bureau Federation came into existence. The objectives of the new organization were to correlate and strengthen State Farm Bureaus; to promote, protect and represent the business, economic, social, and educational interests of the farmers of the nation and to develop agriculture.[17]

The New York State Farm Bureau became an organization endorsed by the farmers of the State, gradually growing to 68,000 members by 1944, with 43,000 members in Home Bureau and 72,000 in 4-H Clubs.

In 1955, the relationship between Farm Bureau and the Extension Service was discontinued. Although the majority of New York's members were happy with the partnership that had successfully operated for over 40 years, groups from some other states were not. The separation of the Farm Bureau and State Extension Service was made mandatory by order of the Secretary of Agriculture, Ezra Taft Benson, in November 1954.[18] The New York State Farm Bureau was formed and incorporated May 18, 1955 officially ending the relationship between the two.

New York Farm Bureau, now totally separated from the Cooperative Extension System, saw its membership drop from 80,000 to 10,000 with the change. Farm Bureau became a non-governmental, volunteer organization financed and controlled by families for the purpose of solving economic and public policy issues challenging the agriculture industry. There was no more financial support from county, state or federal government. Farm Bureau's total support came from its members and the cost of membership had to increase to cover expenses. Membership cost the farmer $20 a year, rather than the minimal cost when it was part of Extension, and many farmers paid $100 up front to provide Farm Bureau with working capital.

As a means of aiding farmers and increasing membership, a group purchasing effort (Safemark) was initiated and the Farm Family Insurance Company was formed to provide an adequate insurance program for farm families. Interest-paying debentures were sold to farmers to provide financing for the new company. Kitchen Konferences were formed where small groups of farmers met on a monthly basis to discuss public policy. Among the groups formed was a Workers Compensation Safety Group, a Farm Bureau Woman's Program and "Youth Power". While all of this was happening, Farm Bureau was working in Albany, Washington and at the county level helping to secure beneficial legislation for agriculture and opposing detrimental legislation.

When Farm Bureau became independent, its first offices were in Ithaca, but in the 1960s new offices were built in Glenmont, near Albany. By the 1970s membership reached 20,000 and today it has nearly 30,000 member families. Currently there are 56 county offices and four regional offices.

During the 1990s, the New York Farm Bureau Agricultural Education Foundation, a charitable education foundation, was formed. Its mission is to inform and educate all New Yorkers regarding agriculture and to increase understanding of agriculture between the farm and the non-farm public. The Foundation distributes an Agricultural School Calendar to 3rd grade classes throughout the State, purchases and distributes agricultural videos to urban schools and visits urban schools to make sure children know how agriculture impacts their daily lives.

Farm Bureau has proven to be an effective voice for farmers and is respected by government at all levels. It continues to be a positive spokesperson for farmers and rural residents throughout New York.

AGWAY
In 1920, the New York State Dairyman's League, the New York State Farm Bureau Association and the New York State Grange voted to form a large, financially strong farmers co-op to meet the growing needs of New York farms. This new co-op was named the Grange League Federation Cooperative (GLF). Ithaca, New York was the headquarters for the cooperative and under the leadership of H. E. Babcock, during the 1920s and 30s, it became a highly respected farmer's cooperative.

---

17   Ibid. p. 12-16
18   Colman, Gould P. *Education and Agriculture* p. 477

In 1964, GLF, Eastern States Farmers Exchange, Inc. (Eastern States) and Pennsylvania Farm Bureau Cooperative Association (PFB) joined together to form a larger and financially stronger farmer's cooperative, which they named Agway. The combined sales of the three cooperatives at the time of the merger were $375 million. Syracuse became the headquarters and a new headquarters building was constructed in DeWitt. Edmund H. Fallon, previously general manager of GLF, became Agway's general manager and J.C. Corwith, a farmer from Long Island, Chairman of the Board.

Agway grew rapidly, reaching out into a variety of agriculture and agriculturally related enterprises. It invested in Mohawk Airlines, which later became US Air, as well as in P&C Food Stores. Agway was influential in the formation of Curtice-Burns, Inc. and Pro-Fac Cooperative, a farmer-owned cooperative whose members sold their fruits and vegetables to Curtice-Burns, who in turn processed and marketed them. In 1980, Agway purchased H.P. Hood, Inc., a Boston based processor of dairy and other food products.

Agway had several divisions that not only supplied farmers with fertilizer, feed and other farm supplies, but also bought products from growers to process and market. They also supplied farmers with insurance, petroleum products and provided credit for leases of large farm equipment. Agway also broadened its markets by establishing corporate stores and franchising representative stores, in addition to the existing local farmer cooperatives. These all sold supplies to the suburbanite and part-time farmer as well as to the full-time farmer. In the 1980s, Agway had over 80,000 farm members and was doing several billion dollars of business each year.

In 1996, Agway sold agricultural products, including fertilizer, feed and seed, in 217 locations, owned 114 retail stores and had 337 franchised retail dealers. Telmark, a subsidiary, financed leases in 27 states.

Agway's emphasis upon size and growth, rather than upon efficiency and profits, began to affect profits in the late 1980s and it lost money in 1990, 1991 and 1992. Agriculture was rapidly changing in the late 1900s and Agway was unable to appropriately change with it. Segments of the firm were gradually sold to obtain cash and Agway finally filed for bankruptcy in 2002. Agway and its predecessor,

GLF, both provided tremendous services to New York agriculture for approximately 80 years.

NEW YORK STATE APPLE GROWERS ASSOCIATION

The New York State Apple Growers Association was formed in 1994 when the New York and New England Apple Institute, which was formed in 1935, dissolved. Its history dates back to the New York State Fruit Growers Association of the early 1900s and later to the Apple Growers of Central and Western New York, formed in 1950. A press release from the Apple Growers of Central and Western New York in 1950, stated it had 1,200 members, producing over six million bushels of apples. In 2009, the New York State Apple Growers Association has 670 members from all corners of the State and is a very active representative of the apple industry.

In the 1950s, legislation was passed in Albany making it possible for agricultural commodity groups to vote for marketing orders, administered by the New York State Department of Agriculture and Markets. In 1959, the New York Apple Marketing Order came into being by the vote of the New York apple growers. This Order made it mandatory for all apple growers to pay into a fund for apple promotion, education and research. Currently, apple growers pay eight cents per hundredweight of apples into a marketing order and two cents per hundredweight into a research order. Every eight years, apple growers vote on the orders, choosing whether to retain or abandon the order along with the funding.

The Association promotes New York State produced and packed apples and apple products through advertising, promotional and educational work with retailers, handlers, consumers, processors and others. They also promote agricultural and market research activities.

NEW YORK STATE HORTICULTURAL SOCIETY

The New York State Horticultural Society (New York StateHS) goes back to 1855. Its mission is to serve the fruit industry of New York through education and promotion, along with protecting the viability of the fruit industry. Traditionally, an annual educational and trade show was held in both Rochester and Kingston, but in 2004 the shows evolved into one annual event that now includes the New York State Vegetable Growers Association. (New York StateVGA). The Empire State Potato Growers and the New York State Berry Growers are also participating. Cornell Cooperative Extension conducts the educational programs for the various fruit and vegetable growers.

The New York StateHS is actively working with the Cornell Experiment Station at Geneva in the development and promotion of new apple varieties adapted to New York State. It is also active in the Council of Agricultural Organizations and works with the other agricultural organizations in the State to carry out programs beneficial to New York agriculture.

EMPIRE STATE POTATO GROWERS

There are approximately 150 commercial potato growers in New York, growing over 20,000 acres of potatoes. This organization is not only important for providing its members with educational and promotional activities, but also for its annual Empire Farm Days. The Empire State Potato Growers, Inc first sponsored an agricultural show in 1931 to bring the potato growers the newest in agricultural techniques and equipment. For many years it was entitled, "Potato Field Days" but in recent years became known as "Empire Field Days". This show has become the largest outdoor agricultural trade show in the northeast with over 600 exhibitors. It is a three-day event with a myriad of activities and events attracting approximately 60,000 visitors each year.

NEW YORK STATE VEGETABLE GROWERS ASSOCIATION

The New York State Vegetable Growers Association was founded in 1911 and is a not-for-profit corporation serving commercial fresh market, storage and processing vegetable growers of the State. The organization provides educational opportunities and resources for its members regarding production, management and marketing, and promotes their products at local, state and national levels.

NEW YORK WINE & GRAPE FOUNDATION

The New York Wine & Grape Foundation was founded in 1985 and currently has 335 members. It promotes the sale of grapes and grape products,

and assists members in the production, processing and marketing of their products. In 2002, it joined with Constellation Brands, Wegman's Food Markets, Inc. and the Rochester Institute of Technology to construct the New York Wine & Culinary Center at Canandaigua. These four organizations had the common goal of creating a gateway for the people of New York and visitors from around the world to experience New York's wine and food industries. The facility opened in 2006 and showcases the food and wine products of the State.

EMPIRE STATE FOREST PRODUCTS ASSOCIATION

This organization dates back to 1906 when the Adirondack Lumber Manufacturers and Shippers Association was formed to keep track of the declining inventories of Adirondack softwood and to promote the emerging science of forestry. Three years later the name of the organization was changed to the Empire State Forest Products Association, to reflect the addition of new members from other regions of the State. Today its membership consists of over 400 businesses and individuals representing all facets of the forestry industry. Its members own and manage 1.2 million acres of New York forests.

NEW YORK FARM VIABILITY INSTITUTE

The New York Farm Viability Institute is a farmer-led non-profit organization that provides grant funding for applied research and outreach education projects that help increase farm profits and provide models for other farms. It grew out of a $993,000 grant from the USDA Rural Business Cooperative Service to Cornell University in 2003. It is an independent organization that is funded by the State and Federal governments in addition to in-kind support from farmers, Cornell University, Cooperative Extension, the SUNY Agricultural Colleges and agribusiness.

The Institute has sponsored a variety of programs, including launching the New York Center for Dairy Excellence and the New York Biomass Energy Alliance. In making project-funding decisions, it prioritizes proposals to obtain the greatest benefit for the dollar in improving the prosperity of New York's agriculture.

Chapter XIV
Marketing of New York's Agricultural Products

For the Native Americans, barter was their method of trading. Agriculture was an important part of their lives, especially for the Iroquois, and corn was commonly traded with other tribes for such things as shells to make wampum. With the arrival of the white man barter continued with corn, squash and beans, although secondary to furs, traded for iron pots, axes and firearms.

In Colonial New York, the early settlers commonly traded the agricultural items they produced with merchants for the necessities they were unable to produce. Dealers normally bartered with the farmers, exchanging goods for farm products, and then sold the agricultural products for cash and restocked with goods for future bartering. There were sales for cash, but it was not the common method of trade in colonial times.

New York State is blessed with one of the world's best harbors as well as having a great many natural waterways. These factors provided a relatively convenient means for farmers to market their products. Between 1714 and 1717, the number of ships clearing the Port of New York each year averaged 64 vessels totaling 4,330 tons. Ships transported grain, flour, pork, bacon, barrel staves, beef butter, candles, hemp and wool to England, the British West Indies, Europe, Africa and the English colonies. There were large shipments of wheat and flour to Boston and to South Carolina. White pine for ship masts came from Upstate forests and pitch pine for tar from Long Island.[1]

England exercised control over what could be shipped from Colonial New York and where it could be shipped until the end of the Revolutionary War. Because of the effects of the war and the need to build trading partners, commerce beyond the borders of the United States developed slowly after the war. Albany became the center for the grain trade after the war, with grain moving north to Canada, east to the New England States and south to New York City. In 1820, there were exports of over 17 million dollars from the Port of New York with 10 million dollars of those exports coming from the soil.[2]

Judge William Cooper of Cooperstown in letters written to William Sampson, a New York City barrister, which were published in Dublin in 1810, expounds some of the advantages New York has for marketing its products. He mentions the importance of the river water routes of the Susquehanna, Tioga, Chenango and Unadilla to Baltimore and points in between: Lake Champlain and the St. Lawrence River to Canada; the lakes Seneca, Cayuga and Onondaga for inland shipping and connections to other waterways. All of these bodies of water including, of course, the Mohawk and Hudson Rivers, were important avenues for the marketing of farm products. In addition, he points out sites for shipping products and describes the rich soils and the fine climate of New York. [3]

1   Kamen, Michael, *Colonial New York* p. 161-171

2   Roberts, Ellis H. *American Commonwealths New York* p. 455-460

3   Cooper, William *A Guide to the Wilderness* p. 22

A load of potatoes and other produce ready to take to market, circa 1914. Straw was put in the bottom of the bobs before loading and the produce covered with blankets to keep it from freezing. Courtesy of the La Fayette Historical Society.

Although the transporting of agricultural products was often by water, in the winter and in certain locations they had to be moved by land. An article in the Albany Gazette of 1804 states that a wagonload of 100 bushels of wheat was pulled by four yokes of oxen the 230 miles from Bloomfield, in Ontario County, to Albany and that a round-trip took 20 days. It further stated that the wheat was purchased in Bloomfield for 62½ cents a bushel and sold in Albany for $2.15½ a bushel.[4]

A note from DeWitt Clinton's Journal in 1810, when he transversed the State studying the practicality of building a canal, stated that cattle are sent in droves to Philadelphia with 2,000 from Seneca County that year. He also mentioned that 200 barrels of beef and pork were shipped to Baltimore from Owego.[5]

Stephen Durfee, a pioneer of Palmyra, in Wayne County, remarked in 1811 that much of the earliest grain produced in that area went to distilleries and to new settlers. He also said that they bartered grain and livestock at the local store for goods that they needed and that sons of farmers often trapped wild animals, in the winter, for their furs as a source of income.[6]

There were many items produced on our farms that we often overlook. A few of the items quoted in the March 25, 1832 *Rochester Prices Current* were ashes, dried apples, bristles, beeswax, clover seed, flax seed, feathers, hops, lard, mutton, mustard seed, dried peaches, salt pork, quills, tallow and buckwheat flour.[7] The bristles and lard were by-products from the hog and the quills and feathers by-products of poultry.

A drover was a common occupation important in marketing up until into the 20th century. Cattle, hogs sheep, geese and turkeys were driven to marketing centers by drovers, who purchased livestock from farmers along their route resulting in sizable flocks or herds by the time they reached their destinations. There were inns at regular points along the way with pens to hold the animals overnight and lodging for the drovers. In 1846, there were an estimated 33,370 head of cattle that passed through Angelica in Allegany County and in 1849, more than 50,000 hogs passed through Fredonia in Chautauqua County on their way to eastern markets.[8]

The completion of the Erie Canal in 1825 and the later opening of other New York canals, dramatically changed access to markets for thousands of New York farms. The waterways of New York now provided access to the markets of the world with

4   Mau, Clayton *The Development of Central and Western New York* p. 131
5   Ibid. p. 160
6   Ibid. p. 98

7   Ibid. p. 277
8   Ibid. p. 391 & 397

substantially less need for transportation on land and without the portages that were common in the past. Grain from Western New York became more valuable to the farmers that produced them, while the value of grain grown by farmers in Eastern New York decreased in value. Although barter continued for another century, it gradually diminished as broader markets opened. Soon after the canals came into use railroads were constructed and highways improved, gradually changing the manner in which agricultural crops were marketed.

Public markets, often termed farmers' markets, have been popular in all parts of the world, for thousands of years. They served as a place where people could obtain food, hear the latest news, be entertained and to see friends. One of the earliest public markets that existed in New York was in what is now part of New York City. A resolution passed on September 13, 1656 stated in part: "Whereas diverse articles such as beef, pork, butter, cheese, turnips, carrots, cabbage and other diverse produce are brought to this city ..., the Director General and the Council do hereby ordain that from this time forward a market will be held here in this city on every Saturday on the beach by or near the house of Mr. Hans Kiersteede according to which everyone who hath anything to buy or to sell shall have to regulate himself".[9]

Two reasons were given for passing the ordinance and they were both farmer-friendly. The first concerned the time wasted by farmers coming to the city to sell their products without buyers knowing when or where the farmers were. The second concern cited was that unsold products often spoiled because of the lack of buyers.

It is likely that almost every city and village of more than a few hundred people had a farmers' market in the days of the horse and buggy. These markets were an important source of income for the local farmers as well as an important means for the

---

9   O'Callaghan, Edmund Bailey *From the Laws and Ordinances of New Netherland* p. 251

Farmers lined up with their tobacco crop waiting to have it weighed and loaded on railroad cars, circa 1897. This was a once-a-year occasion with the farmer receiving pay for his year's work at this time. Courtesy of the Museum at Shacksboro Schoolhouse, Baldwinsville, NY.

New West Washington Market in New York City from an 1888 photograph in Harper's Weekly. Note that most of the produce is packed in wooden barrels, the dominant container at the time.

Niagara Frontier Food Terminal, circa 1932. This was, initially, a 60-acre wholesale market with a 10-acre farmers' market across the street. Courtesy of Cornell University.

Farmers' market adjacent to the Erie Canal in Syracuse, circa 1895.

village residents to obtain food. Syracuse was just a small village, in 1835, when it constucted a special building to house its farmers' market. It wasn't profitable but a later open public market in the center of Syracuse, along the Erie Canal, was very successful. When plans were made to reroute the Canal away from the city, the market moved, in 1899, a few blocks away. This market also prospered until it was moved to a large, new regional market facility, named the Central New York Regional Market, on the edge of the City, in 1938.

There were several regional markets established, in the same time period, in central metropolitan areas in the State, including Albany, Utica, Buffalo, Rochester and New York City. Farmers, who made retail sales to consumers as well as selling wholesale to retail stores and hucksters, and wholesale dealers, both used these markets. A 1940 summary of the CNY Regional Market relates that the total sales for the year were $1,670,698, with vegetables comprising 50% of the dollar value, fruit 15%, poultry and eggs 30% and miscellaneous items 5%. There were 297 different growers, bringing 48,947 loads of produce from 10 counties. Wholesale buyers came to the market from both Central and Northern New York. There were a large number of farmers selling live poultry directly to the consumer. Some of the poultry customers slaughtered their purchases at home and some took the poultry to a nearby building where they paid a custom processor to kill and process the birds.

There were very few chain stores, prior to the 1950s, and it was common practice for the proprietors of retail stores, as well as hucksters who traveled from house-to-house in the city selling their fruits and vegetables, to arrive at the market early each morning to purchase their needs for the day. When the supermarket chains replaced the corner stores and the hucksters, local farmers' markets suffered from both lack of wholesale business and retail

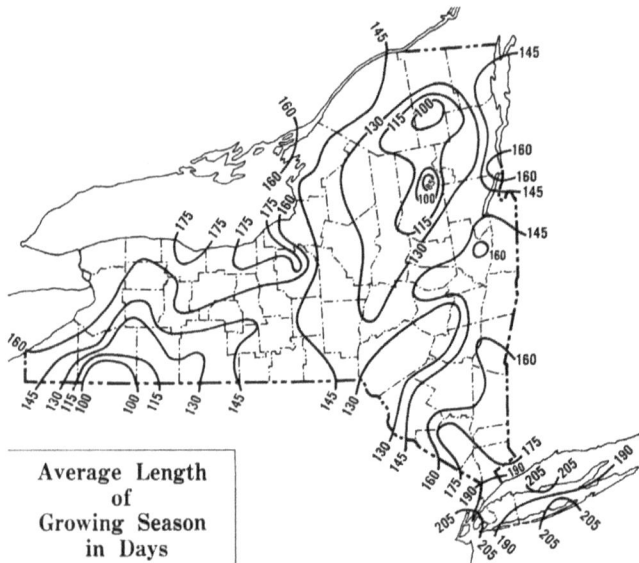

Average Length
of
Growing Season
in Days

This map shows the average length of the growing season in various areas of New York in 1919, which would be quite similar in 2009. Note that the average length of the New York growing season varies from 100 days to 205 days. This is just one of the factors that contributes to the great diversity of crops grown in New York. Courtesy of New York State Archives.

sales because the consumer now found a vast array of fresh produce, from all over the world, available at the supermarket. Many farmers' markets ceased to operate and many small farmers lost a market for their products because the quantity produced by each farmer was too small to be of interest to a supermarket chain buyer.

Farmers' markets started a recovery in the 1970s with increased consumer interest in eating fresh fruits and vegetables to improve their health. Some consumers also had concerns about purchasing products grown hundreds or thousands of miles away, because of the perception of an excessive use of pesticides in crop production and their feeling that produce grown closer to home would be fresher and perhaps safer. The higher transportation costs and the carbon effect of food traveling long distances, encourage farmers to produce locally, just as they encourage consumers to purchase food grown locally.

In 1976, the Council of the Environment of New York City established the Greenmarket program, which provided regional small family farmers with opportunities to sell their fruit, vegetables and other farm products in open-air markets in the city. The most famous of these is the Union Square Green Market, held four days a week, and attracting

250,000 customers weekly, purchasing a vast variety of fruits and vegetables.

New York State's government, operating through the Department of Agriculture and Markets, has been working diligently, in recent years, to encourage healthy eating by New York consumers through the purchase of fruits and vegetables from local farmers offering their products at farmers' markets. Beginning in 2000, a Markets Grant Program was initiated, which offers matching grants of up to $25,000 for improvements to farmers' markets. Awards of almost $1 million have been made to 63 markets since the program began. In 2008, eligible seniors and participants in the Women, Infants and Children (WIC) program redeemed over $5.6 million in Farmers' Market Nutrition Program checks to purchase fresh, locally-grown fruits and vegetables at farmers' markets across the State. There were 975 New York farmers at 400 farmers' markets across the State participating in the program.

Another method of marketing has been recently established at Penn Yan. There are a large number of Amish farmers in the area and a wholesale produce auction has been established.

Yet another form of direct marketing that has been increasing in recent years is retail sale of agricultural products on the farm. When the automobile came into common use in the 1920s, farms became closer to the suburban dweller in terms of time, and many farmers saw the opportunity to grow products for this market. When the large supermarket, with its vast array of products, arrived after World War II, and the number of farmers declined dramatically during the same period, the number of farm stands decreased. There has been an increase in the number of on-the-farm retail outlets during the past two decades for the same reasons that farmers' markets have increased. Additionally, New York's Amish and Mennonite farmer population has increased and these people have traditionally marketed part of what they produce at their farm stands. There are also a number of New York farms selling the "farm experience". They offer entertainment, as well as sales, by providing petting zoos, hayrides and other attractions to bring the city and suburban dweller to the farm. Sometimes these customers come to experience the pleasure of harvesting the food they purchase by picking berries and other fruit.

MARKETING OF DAIRY PRODUCTS

Dairy products have been a major part of New York agriculture since the early settlement of the State by the Europeans. Information concerning the marketing of milk products could fill books, but these paragraphs provide a brief overview of this important topic.

Milk, until after the middle of the 19th century, was marketed in the form of butter or cheese, both produced in the farm home. Fresh milk was not consumed as commonly as today and was normally available only to families with a milking cow or from a dairy farmer peddling his milk in a neighboring village.

Cheese production in the homes peaked around 1865 because of the desire by markets for a consistent product produced in large quantities. Cheese factories began to appear at scattered points around the State, especially in towns with railroad depots where cheese could be easily shipped to distant markets. With the advent of cheese factories many farms discontinued producing cheese and instead sold their milk to cheese factories.

Butter production in the farm home continued to increase until the late 1800s when creameries were constructed for the production of butter and the shipment of fluid milk to market. Butter production in the home then rapidly diminished, but continued, to a limited extent, into the 1930s.

Occasionally those dairy farmers close to a village or city, chose to sell their milk directly to the consumer by transporting it with horse and wagon to the village and dipping the milk out of a can directly into the customer's container. For example, George Weeks, of Skaneateles, peddled milk to his neighbors in 1856 for 4 cents a quart. As the State's population increased, more farmers began to market their milk in this manner. Some farms continued peddling their milk well into the 20th century, updating to bottles and pasteurization as these practices became available.

The large majority of farmers did not have access to retail markets, however, and sold their milk to cheese factories or creameries. An example of farmers banding together to sell their milk directly to the consumer occurred near Syracuse in 1872 when the Onondaga Milk Association was formed to supply the city with, what they advertised as "good and pure milk". At that period of time, milk quality was often poor because of lack of sanitation on the farm and inadequate refrigeration. The Association had 46 stockholders with a total of 1,116 cows. In 1875, it peddled 18 routes and by 1876, had 37 routes. In 1876, it received 730,498 gallons of milk and made 14,186 pounds of butter and 46,365 pounds of cheese. Although not a cooperative, it functioned similarly to a cooperative as the farmer-stockholders benefited, based upon the profits.

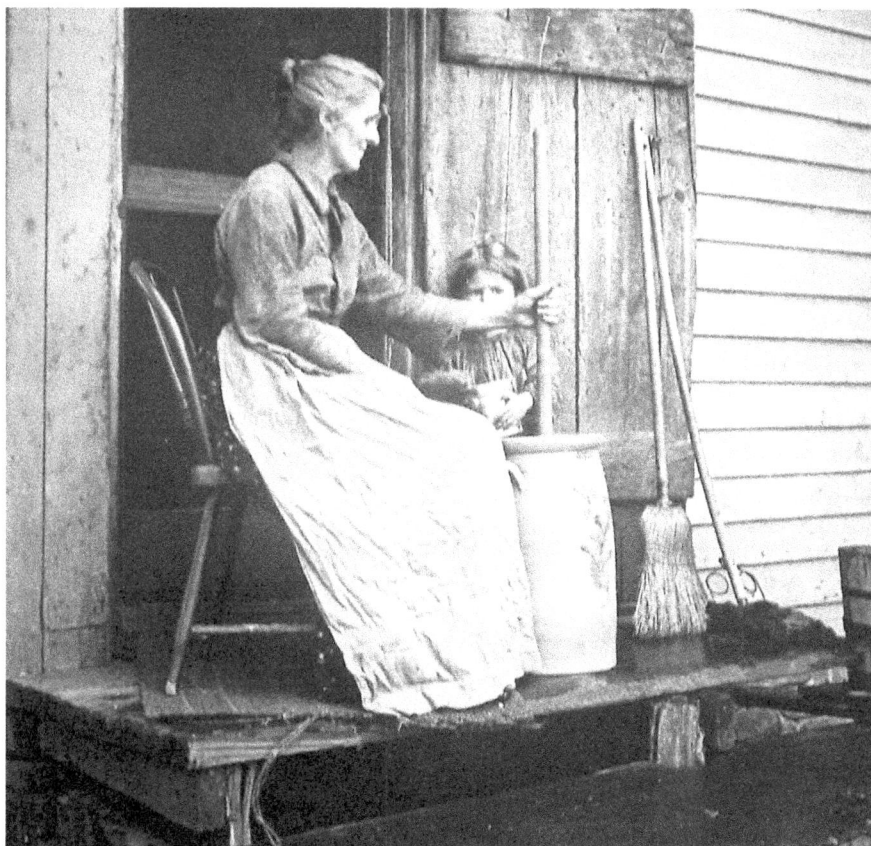

Churning butter on the back step.

A horse-drawn milk wagon ready to deliver milk to village customers, who came out to the street and milk was dipped from the can to fill their containers, circa 1910. Courtesy of the Town of Onondaga Historical Society.

Milk being delivered by farmers to a cheese factory, courtesy of the Town of Onondaga Historical Society.

THE BIG CHEESE, 11,500 POUNDS, ON IT'S WAY TO
SAN FRANCISCO, AS A PART OF THE N.Y. STATE EXHIBIT.
MADE IN LEWIS CO. PHOTO TAKEN JAN. 27. 1915.

The 11,500 pound "Big Cheese" from Lowville in Lewis County, on its way to San Francisco as a part of the New York State exhibit in 1915, courtesy of the New York Museum of Cheese.

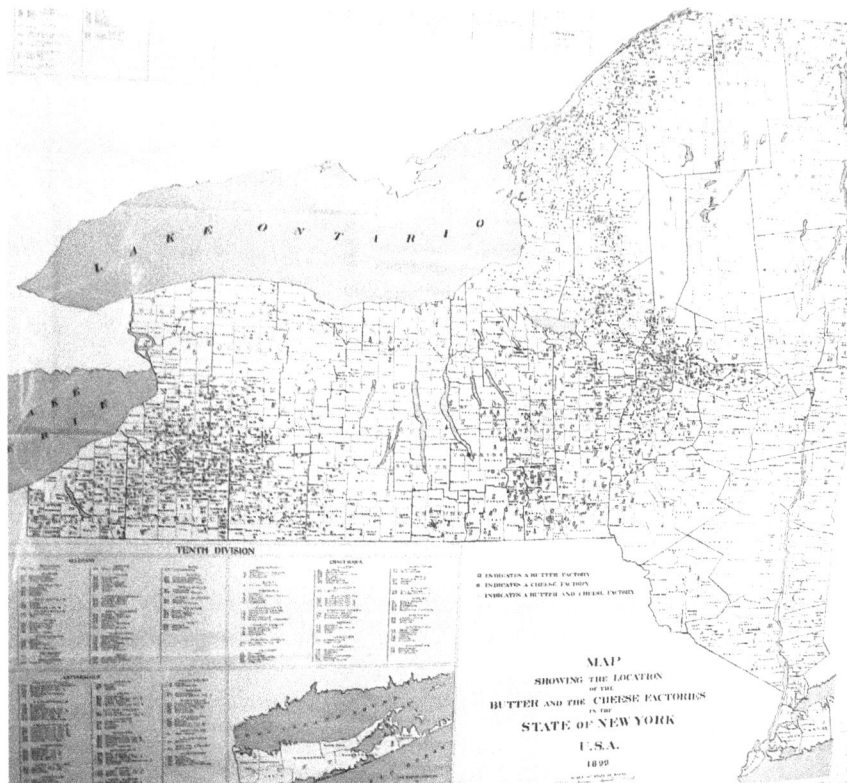

Map showing the locations of New York butter and cheese factories in 1899. Each dot represents a factory and there were hundreds!

Throughout New York's history, farmers have often been paid less for their milk than the cost of its production. This has been caused by both over-supply and milk dealers working one farmer against another in an attempt to purchase it at the lowest possible price. Because of this, there were movements, going back to the late 1800s, to establish cooperatives in an effort to receive a fair price for milk. Dairylea, originally formed under the name of the Dairymen's League, Inc., a cooperative that has been very successful, had its beginning in the Orange County Pomona Grange in 1906. Farmers there had attempted to form a bargaining organization to improve the price of milk since 1898, but without success. In 1907, dairymen in Orange, Sullivan and Ulster Counties, along with dairymen in Sussex

A Dairymen's League milk delivery wagon, circa 1922.

County, New Jersey, formed what we now know as Dairylea. The territory covered by the new organization was extended beyond the initial area, as rapidly as possible, and grew to include dairymen from New York, Pennsylvania and New Jersey. By 1910, it had over 10,000 members, owning 190,000 cows.[10]

During this time period, country milk stations were used in conjunction with the railways. Farmers brought their milk to railway collection points, where the milk was cooled and then shipped as far as several hundred miles to metropolitan areas for processing and consumption. Milk coming to metropolitan centers from many distant points made it necessary for farmers to join together from large geographic areas for bargaining success.

In 1916, milk dealers refused to recognize the authority of the Dairymen's League, so the League called a milk strike that lasted 11 days, until the dealers agreed to the League's pricing terms. Following this success, farmers were eager to join the League and the membership grew, in less than a year, to 42,000, representing over 500,000 cows. In 1919, there was a second break with the milk

dealers, which lasted 18 days before the dealers capitulated to the League. Subsequently the League adopted a pooling plan providing payments to be made to farmers on the 25th of each month. This pooling plan went into effect in 1921 and the new organization became the Dairymen's League Co-operative Association. The new organization soon was purchasing and building milk plants and manufacturing facilities to process their members' milk. In less than a year the League had a total of 78 plants, including 11 milk-condensing plants.[11]

The passage of the Capper-Volstead Act in 1922 clarified the legal authority of cooperative associations for "purposes of producing, handling and marketing farm products". This legislation gave cooperatives the green light to pursue marketing, both legally and aggressively. This law has been referred to as the "Magna Carta" for farmers.[12]

In 1923, the Dairymen's League Co-operative Association began marketing its own products under the brand name Dairylea. Membership as well as markets increased, but by 1932 farmers were averaging less than one-cent a pound for their milk.

10 Alexander, L. Ray *History of the New York State Grange 1873-1933* p. 128-129

11 Ibid. p. 129-131
12 Dairylea *Dairylea 100 Years of Service* p. 10

The Pitcher Committee was formed by the New York State legislature to look into the situation and in 1933 it passed the Milk Control Law, which established a milk authority to oversee the production and distribution of milk. This ineffective law was later declared unconstitutional. In 1937, however, the New York legislature passed the Rogers-Allen Law giving bona fide cooperatives the right to function together in establishing and enforcing prices and marketing programs. From this legislation the Metropolitan Cooperative Milk Producers Bargaining Agency was born. Within a year the milk price jumped from $1 to $2 per hundredweight as 100 cooperative groups, representing 50,000 members, worked together.[13]

Prior to the passage of the Rogers-Allen Law, the cost of milk production was greater than the price most farmers received. There had been numerous milk strikes by desperate farmers in attempts to receive a fair price for their milk, but success had been limited. Higher prices for milk brought surpluses that could not be utilized in the marketplace and milk was sold at distressed prices or was destroyed. It had been common practice for milk dealers to work one farmer against another in order to purchase milk from the farmer willing to sell at the lowest price. Finally, the New York Market Order was approved in 1938, which stated that all milk handlers must purchase milk for the same price for the same use and that all producers shall receive the same price regardless of use, except for transportation costs related to varying distances from markets. This Order stopped dealers from auctioning the price down to the lowest bidder at both the federal and the state levels. The Order was subsequently declared unconstitutional, but in 1939 the United States Supreme Court reversed the decision. Later, radical elements stirred discontent among some farmers with a violent milk strike in August of 1939. The strike was settled in the offices of New York City Mayor Fiorello La Guardia resulting in an increase in the milk price.[14]

An additional positive result from this Order was that it removed any concerns the farmer had regarding the marketing of his milk and permitted him to concentrate on production, which resulted in greater production efficiency. Supply and demand became the regulators of the price of milk once the Order came into existence.

13 Dairylea *Dairylea 100 Years of Service* p. 19-20

14 Ibid. p. 20-23

Creamery and milk factory at Fabius in Onondaga County, circa 1900. Courtesy of Bill and Joanne Casey.

Sheffield Farms milk factory in Schoharie County in 1913. Courtesy of New York State Archives.

Farmers increased milk production during World War II in response to the government's urging, but after the war ended, were hit with low prices from surpluses. The 1950's brought about the beginning of the end of the 10-gallon milk can used by the farmer for 50 years. It was replaced by bulk refrigerated milk tanks on the farm and bulk milk hauling. This change was a great improvement but was also "the straw that broke the camel's back" for many marginal dairy farmers. Dairy cow numbers per farm, which had been increasing during the war years, needed to increase further, on many farms, to make the installation of a bulk milk tank practical.

Onondaga County provides an example of the consolidation that has occurred in the dairy processing industry throughout the State. In the 1950s there were 23 dairy processors processing and distributing milk in the County. At that time, almost every village was supplied by more than one local dairy processor. Byrne Dairy, founded in 1933 by Matthew Byrne, and still owned and operated by the Byrne family, is an example of a dairy that survived. They bottled and distributed milk in the Syracuse area, but in recent years have expanded into much of the Northeast. In 1952, they opened their first retail store and now have 55. In 2003, they constructed a 35,000 square foot plant in East Syracuse for manufacturing and packaging. Byrne Dairy has also teamed with the Organic Valley Cooperative to provide organic milk for Byrne Dairy deliveries.

The Dairymen's League is another example of the changes needed to survive in the milk marketing industry. In 1969, it renamed itself Dairylea

Cooperative Inc. It expanded its product line beyond whole milk to include low-fat milk, ice cream, cheese, cottage cheese, ultrapasteurized products, milk powder, butter, yogurt and more.[15] A greater variety of dairy products are being made available for the consumer with each passing decade.

Milk prices move up and down with supply and demand and in 1959, when there were low prices with a surplus of milk, an attempt was made to minimize this problem. A number of dairy farmers formed the O-AT-KA Cooperative to help balance the supply of milk. Their plant produced butter and milk powder in an effort to take surplus, price-depressing milk off the market. Today, 90% of this plant is owned by Upstate Niagara and 10% by Dairy Farmers of America. O-AT-KA makes a large variety of value-added milk products and has routinely returned annual profits to its member-owners.

Difficult financial times struck Dairylea again in the 1970s, prior to when Clyde Rutherford was elected president. At that time, Bobby L. Hall was successfully managing the O-AT-KA Cooperative's butter and milk powder processing plant at Batavia, when Rutherford and the Dairylea Board asked him to help Dairylea overcome its financial difficulties. Subsequently, Hall, Rutherford and the Dairylea Board brought in Rick Smith, an attorney, to help untangle a number of legal problems. They decided to sell all of the firm's operational facilities, which was completed in the late 1980s. Prosperity gradually came back to the company. Bobby Hall moved on to manage processing plants, which were his real interest, and Rick Smith became the CEO of Dairylea. Success was evident, as by 1999 Dairylea had become the fifth largest cooperative in the United States.

There have been other successful milk cooperatives in New York in addition to Dairylea. The formation of Upstate Farms Cooperative, Inc. in 1971 was the result of the merger of several independent cooperatives: Arcade Farms Cooperative, Inc.; Arcadia Cooperative Marketing Association, Inc.; Buffalo Milk Producers Cooperative Association, Inc.; Collins Producers Cooperative, Inc.; Community Cooperative Marketing Association, Inc.; Genesee Milk Producers Cooperative, Inc.; Hollisville Milk Producers Cooperative, Inc.; Rochester Guernsey

---

15 Dairylea *Dairylea 100 Years of Service* p. 47

Producers Cooperative, Inc.; Rochester Independent Milk Producers Cooperative, Inc.; and Western New York Milk Producers Cooperative Association, Inc. The Eastern Milk Producers Cooperative, headquartered in Syracuse, was a bargaining agency for farmers. Eastern merged with Milk Marketing Inc. in 1995 and later Milk Marketing Inc. consolidated with three other milk associations to form Dairy Farmers of America.

When, in 1994, Upstate Farms was having financial difficulties, it also called upon Bobby Hall to help. With Chairman Don Rudolf, Vice Chairman Dan Wolf, the Board of Directors and Bobby Hall working together, Upstate Farms was turned into a profitable cooperative. Upstate built a cultured milk plant near Buffalo for the production of yogurt, sour cream, cheese dip and cottage cheese in 2004, and in 2006 merged with Niagara Milk Marketing Cooperative to form the Upstate Niagara Cooperative. Today, Upstate Niagara is very successful and is marketing its products in 49 states.

In 1997, the Northeast Dairy Herd Improvement Association (NeDHIA), later renamed Dairy One, partnered with Dairylea to pool resources and maximize opportunities for farmers on herd health, milk quality, forage testing and software services. Cooperative ventures of dairy cooperatives continued with the partnering of Dairylea with the Dairy Farmers of America in 1999, resulting in the formation of Dairy Marketing Services (DMS). The goal of DMS was to provide a superior market, with a profitable return, for members' milk by combining the resources of the largest cooperatives in the Northeast with the largest cooperative in the United States. A major advantage of this cooperation is the more efficient collection of milk from member farms. A single milk truck can pick up milk from the producers of several different organizations, removing the need for multiple trucks to travel the same roads to pick up milk for several cooperatives.

In 1998, the merger of four of the country's dairy cooperatives created Dairy Farmers of America (DFA), the largest cooperative in the country. In 2001, Dairylea also became a cooperative member of DFA and in 2003, Rick Smith, COO of Dairylea, became president of DFA.

The continual merger of small milk cooperatives into larger ones has followed the path of other American industries seeking to gain efficiency, lower costs and provide a greater return. Indirectly the consumer, as well as the farmer, is the beneficiary, because a good portion of the lower costs of marketing and processing are passed along to the consumer.

Sunoco's ethanol plant at Fulton at the site of the former Miller Brewery. It is expected to be fully operational in 2010 and produce 100 million gallons of ethanol a year. The plant is anticipated to use over 30 million bushels of corn annually from New York and other states.

153

The numbers of both New York dairy processors and dairy farmers have dramatically decreased during the past 60 years. The quantity of milk produced and consumed in the State has continued to increase with the population. The price of milk received by the farmer continues to fluctuate with supply and demand. In past years the fluctuations in price were caused by supply and demand in smaller geographic areas, but now supply and demand, from countries on the other side of the world, affect prices received by our New York dairy farmers.

Chapter XV
Agricultural Education, Research and Extension

Until the late 1800s, neither the State or Federal government provided funds for agricultural research or teaching. Some educated landowners experimented and exchanged information with each other or read of agricultural practices in Europe. Agricultural publications began to appear, during the early 1800s, providing an exchange of ideas among farmers. For several decades there were numerous attempts by individuals and various agricultural groups encouraging the New York State legislature to appropriate money for an agricultural school. Jesse Buel, secretary of the Ag Society, proposed the plan for an agricultural college in 1833 and the Society made eight additional attempts before a College of Agriculture was established in 1865 at Cornell University, three years after the Morrill Act of 1862 became law.

There had been earlier, feeble attempts to provide agricultural education in the State. Back in 1754, Kings College, now Columbia University, provided instruction in husbandry and commerce. In 1792, the State Legislature granted funds to Columbia University for the establishment of professorships in natural history, chemistry and agriculture. Agricultural interests wanted much more than a few professorships in New York City. They wanted a full-fledged agricultural college with a working farm for both research and experience for the students. In 1839, 80 petitions with 6,000 signatures for a school of agriculture were presented to the State Assembly. A statement accompanying the petitions expressed surprise that there was no school in the State where the science of agriculture is taught though

it is "a business which occupies eight-tenths of our population".[1]

In 1856, the State did make provision for a loan of $40,000, for 21 years, for an agricultural college and farm at Ovid on the condition a similar fund was raised by the institution. The college initially appeared to be a success, but the Civil War came along and the entire student body, including the college's president, went off to war. The school closed and never reopened, with its land and buildings eventually used by the State for another purpose. The People's College of Montour Falls was incorporated in 1853 and was later offered land through the Morrill Act of 1862. The Morrill Act provided a gift of public lands to the states, for the establishment in each state "of at least one college where the leading object shall be to teach such branches of learning as are related to agriculture and the mechanic arts". By 1865, the People's College had not opened its doors, was unable to meet the stipulated conditions of the Act and the land grant subsequently went to Cornell.[2]

Ezra Cornell had initially been interested in splitting the Land Grant between the People's College and the Agricultural College at Ovid. At this time Cornell was serving in the New York State Senate and clashed with one of his colleagues, Andrew D. White, who had his own vision of a great new-type of University and strongly advocated keeping the Land Grant fund intact. New York was the most

---

1   Betten, Cornelius *A Century's Progress in Agricultural Education*
2   Ibid.

populous state in the Union, at that time, and since the amount of land script was based on population, New York received script for 990,000 acres. After much discussion, Ezra Cornell was won over to White's viewpoint and offered to endow a new university with $500,000 of his own money if this new institution received the entire proceeds of New York's share of the land-script from the Morrill Act.[3]

Ezra Cornell, founder of Cornell University, circa 1867.

Ezra Cornell introduced legislation containing his plan for a new university, and nearly all of the colleges in the State descended upon Albany to oppose the legislation. After substantial discussion and debate the entire Land Grant went to Cornell's university. Cornell University was chartered in 1865 and opened its doors to its first students in October of 1867, with Andrew D. White as its first President. One of the stipulations of the legislative decision in Albany was that Cornell University was to educate, tuition-free, one student from each of the State assembly districts. Another stipulation was that the Governor, Lieutenant Governor, the Speaker of the Assembly and the President of the New York State Agricultural Society were to be ex-officio members of the Board of Trustees.[4]

Ezra Cornell, who made his fortune in the Western Union Telegraph, was astute in making certain that Cornell received the greatest value possible from

its Land Grant script. He personally selected, for Cornell, what he considered to be the best lands in Michigan and Wisconsin. There was a large surplus of land on the market, resulting in low land prices, since most states immediately sold the lands they received from their script. Ezra Cornell insisted that Cornell University retain its land until the price appreciated and received a great deal of criticism for doing this. It proved to be a very wise decision as Cornell was able to obtain approximately $3,000,000 for its land, far more than any university from the other states.

The 1876 report by the Commissioner of Agriculture of the US Department of Agriculture indicates that the US government was monitoring the Land Grant colleges' use of the funds received from the sales of their respective lands. A few items of interest in the report included: 1) Cornell has a farm of 150 acres valued at $22,000, 2) experiments were conducted on the value of oats as a soiling crop, 3) the effects of a variety of commercial fertilizers including gypsum and ashes on crops, 4) a comparison of planting soaked and unsoaked seed, 5) the proper number of cornstalks for a hill and 6) the value of crossing the common cow with certain purebreds. It also reported that failures in farming resulted not so much from poor soil as from poor culture. Additionally, the report stated that during 1876 Cornell received $35,000 from the proceeds of its congressional land-script and sold 17,447 acres of the land at an average of $4.65 an acre with 375,000 acres remaining unsold.

At the time Cornell University was founded, there were few trained agriculturalists in America and the first attempts to secure an appropriate professor of agriculture were met with failure. In 1873, Cornell University appointed Isaac P. Roberts as Assistant Professor of Agriculture. Roberts had grown up in New York, moved to Iowa to manage farms and was so highly respected that he was named Professor of Agriculture at Iowa State College. He proved to be a successful leader and teacher at Cornell and was appointed Director of the Agricultural College in 1878.[5] Some years prior to Roberts' arrival at Cornell University, the University had been successful in securing James Law, a noted Scottish veterinarian, as Professor of Veterinary Medicine. As Andrew D.

3   Carron, Malcolm S.J. *The Contract Colleges of Cornell University* p. 11-13
4   Ibid. p. 13-14

5   Hendrick, Ulysses P. *A History of Agriculture in the State of New York* p. 422-424

Cornell students on what is now the Arts quad, circa 1872. Courtesy of Cornell University.

White had embarked on a trip to Europe for the purchase of equipment for the new University, Ezra Cornell shouted to him from the wharf, "bring back a horse doctor".[6]

In the early years of the University it proved not only difficult to obtain qualified professors of agriculture, but it was also difficult to attract qualified students interested in studying agriculture. Many people scoffed at the value of a college education in agriculture. In 1873, there were only seven agriculture students at Cornell so in 1874, as a means of encouraging enrollment, the University stated that every student who pursued a course in agriculture would receive free tuition. This action appeared to be successful as there were 18 agriculture students that year and the number increased to 42 in 1877. The lack of interest in the College, at that time, is magnified by the fact that the total University enrollment, during each of the years from 1873-1878, was over 500.[7]

The "extension" of scientific knowledge directly to the farmer was initiated almost immediately with Robert's arrival at Cornell as Assistant Professor.

Robert's goal was to bridge the gap between practice and science. During Robert's career he wrote 1,400 articles for the agricultural press and with the establishment of the Cornell Experiment Station in 1879, he helped write bulletins of agricultural information beneficial to New York's farmers. Since money was not provided for their printing and mailing, Jennie McGraw Fiske financed the bulletins, from 1879-1885.[8]

In 1880, State legislation provided for a New York State Agricultural Experiment Station. Consideration was given to many sites and finally, in 1882, the New York State Agricultural Experiment Station was established in Geneva. An experiment station had been founded at Cornell in 1879, at the urging of George Caldwell, a professor of agricultural chemistry, in anticipation of funding by the Legislature. When State funding failed to materialize, funding for the experiment station was provided by donations from professors and prominent farmers, along with income from food sales, until the University provided funding in 1881. After the passage of the Hatch Act, this experiment station became known as the Cornell University Agricultural Experiment Station.

6   Carron, Malcolm S.J. *The Contract Colleges of Cornell University* p. 16

7   Coleman, Gould P, *Education and Agriculture* p. 68-69

8   Smith, Ruby Green *The People's Colleges* p. 14-16

Liberty Hyde Bailey in front of the College of Agriculture Experiment Station at Cornell, circa 1900. Courtesy of Cornell University.

The passage of the Hatch Act by the United States Legislature, in 1886, for the establishment of state experiment stations, in connection with state agricultural colleges, was the impetus for Cornell Trustees to organize an agricultural experiment station at Cornell in 1887. The Hatch Act provided $15,000 annually to each state with an experiment station, with Cornell receiving the money in 1888 for use in the Cornell University Agricultural Experiment Station.

Cornell had received no State aid until 1893, when money was appropriated for a dairy husbandry building. The following year, the Legislature made an appropriation for the establishment of the New York State College of Veterinary Medicine at Cornell. Also, in 1894, the fruit growers of New York, under the leadership of Liberty Hyde Bailey, obtained an appropriation of $8,000 for extension work in agriculture. Bailey made a point of taking research and modern methods directly to the farmers of New York through bulletins and extension programs.

In 1903, Isaac P. Roberts retired as Director of the Agricultural College at Cornell and Liberty Hyde Bailey was appointed to succeed him. Bailey had come from Michigan State Agricultural College to Cornell in 1888. At the time Bailey became Director, the College of Agriculture had a record enrollment of 247 students and had even turned away 50 applicants for its winter dairy course. The College was very crowded with little building space.

Obtaining the New York State College of Agriculture at Cornell was the culmination of a long struggle. President Schurman, in his inaugural address as President of Cornell in 1892, summoned the State to provide a $200,000 building for agriculture, but Cornell only received $50,000 for a dairy building. Bailey suggested taking the message of the need for the new agriculture building to the countrymen and develop a lobby of the farmers of the State. Isaac Roberts, Director of the College of Agriculture, said to Bailey, "Go to it, boy". The following year Bailey traveled 28,000 miles around the State, visiting Granges and farmers' gatherings, talking to them of the need for a great agricultural school. Because of this "grassroots" support and Bailey's persistence, the New York State College of Agriculture came to Cornell.[9] A number of other New York colleges wanted to have the New York State College of Agriculture on their campus, but Cornell University was recognized and respected by both farmers and the State's agricultural organizations for its active work with farmers, and their endorsement brought the New York State College of Agriculture to Cornell in 1904. When the Legislature designated Cornell University as the site for the State College of Agriculture, it made an appropriation of $250,000, of which not more than $125,000 was to be used for the main agricultural building.[10]

---

9   Bishop, Morris *A History of Cornell* p. 365-369
10   Carron, Malcolm S.J. *The Contract Colleges of Cornell University* p. 70-71

For half a century, during a period of time when New York was the number one agricultural state in the country, New York's farmers moved forward with almost no support from the State, for agricultural education and research. It was over a century after New York was formed before the State recognized the need for an experiment station and a New York State College of Agriculture. Fortunately the State eventually realized the importance of agricultural research and education giving our New York farmers some badly needed support. Although Andrew D. White was frustrated in his early efforts to develop a strong agricultural program at Cornell, by the late 1800s and early 1900s, under the leadership of Roberts, Bailey and others, Cornell was providing extensive programs in research and education for our New York farmers.

Cornell University has a rich history in extending knowledge, gained in research, to the residents of New York. Initially extension was directed toward farmers, but gradually moved to include farm women, rural school children and rural communities and then, on to serve the total population of the State. Cornell University's first steps in extension occurred in 1869 when Professors Caldwell, Law and Prentiss gave lectures around the State to groups of farmers. A significant additional effort, in extending research to farmers, was due to the efforts of Liberty Hyde Bailey and J.S. Woodward, an editorial writer for the *Rural New-Yorker*, along with the blessing of Cornell's President Adams, when a "Farmers' Institute" was held for three days at Cornell in 1886. It proved to be both popular and successful, continuing the next year in three different locations in the State.[11] The New York State Agricultural Society appropriated $1,050 for holding the three Institutes, but actually five were held. In 1887, the State provided the Ag Society with $6,000 to hold Institutes and 20 were held in various locations around the State. The Farmers' Institutes were established to give farmers the results of scientific research and translate it into agricultural practice. These institutes continued for over half a century with perhaps the most notable one being known as "Farm and Home Week", first held in 1909 at Cornell for a week in late winter.

ERIE COUNTY
FARMERS' INSTITUTE

will be held in the Methodist church,

ALDEN, N.Y.,

beginning at 10 A. M.,

Saturday, Nov. 26, '92.

PROGRAMME:

Prayer by Rev. Wm. Hartwell.
Music.
Address of Welcome by Ira B. Hawthorne.
Response by Henry Adams of Marilla.
Music.
Question Box.

Afternoon Session, 1.30 o'clock.

Music.
Paper on Poultry Raising by W. C. Tucker of Alden.
Music.
"Farm Economy from a Woman's Standpoint," Mrs. Henry Ewell of Alden.
Music.
"Hygiene," Dr. C. A. Tyler of Alden.
Music.
Paper on "Extermination of Noxious Weeds," Henry Adams of Marilla.
Miscellaneous business.
Music.
"Silos," Mr Zurbrick of Clarence.
Music.
"Benefits of Natural Gas," Ambrose Pattison.
"Maple Sugar," W. C. Field.
"America."

A New England Dinner will be served by the ladies of the Alden Cemetery Association for 25 cents each.

C. M. BAKER, Sec'y.

Advertisement for an 1892 Farmers' Institute to learn the latest in farming techniques. Note that the ladies are serving a New England dinner for 25 cents a person! Courtesy of the New York History Museum.

The booklet, *New York State Fair 1841-1917*, has a section extolling the merits of New York as an outstanding agricultural state. It states that, in the late 1800s, four cents per capita was spent for agricultural education and that by 1917 it had increased by six times to over twenty-four cents per capita. One important aspect of the increased attention to agriculture was the creation of the State Department of Agriculture in 1893. New York had, for years, been dominated by its agricultural industry and with the

---

11   Hedrick, Ulysses Prentiss *A History of Agriculture in the State of New York* p. 426-428

A Farm Bureau agent with his Buick Roadster visiting a farmer, but being pulled out of mud by horses and a wagon with milk cans on back, circa 1920. Courtesy of Cornell Cooperative Extension of Onondaga County.

demonstration schools of three days each and the annual Farmers' Week at Cornell with 2,654 attendees. The report also stated that in the 1919-1920 school year, there were 1,216 regular students, 89 special students, 396 winter course students, 530 summer school students and 229 graduate students in the College of Agriculture.

Home Economics had its beginnings in Cornell's Agricultural College when Bailey hired Martha Van Rensselaer to establish a reading course for farm women. By 1907, her work had developed into a full-fledged academic department within the New York State College of Agriculture entitled, Home Economics. Home Economics came into its own when the New York State College of Home Economics was established in 1925.[12]

increasing growth of manufacturing and commerce the state leadership recognized that it was time to direct some resources toward strengthening its agricultural segment.

The Federal government entered into the realm of continuing extension education programs for rural farm families in 1914 with the passage of the Smith-Lever Act. This law established a partnership between the US Department of Agriculture and the Land Grant Universities by establishing the Cooperative Extension System and providing a base of federal funds for its use. It brought about the establishment of the Extension Service in New York, administered by Cornell University. The Smith-Lever Act was followed by the Smith-Hughes Act of 1917, which promoted vocational agriculture and the George-Reed Act of 1929 that expanded the funding for vocational education in home economics and agriculture. **Note:** In this book's chapter, "Agricultural Organizations" under the heading of Farm Bureau, there is further information concerning Cooperative Extension as well as later in this chapter.

The 33rd Annual Report of the New York State College of Agriculture at Cornell University and the Agricultural Experiment Station to New York State in 1920, by Dean and Director A.R. Mann, lists 379 Farmers' Institutes, 3 farmers' field days, 49 farm

There have been many challenges turned into notable and worthy accomplishments by the College of Agriculture that would take a number of books to explore. An example of rising to the need, regardless of the difficulty, took place during World War II. The US government encouraged increased food production during the War and one means to accomplish this was with the establishment of Victory Gardens. Cooperative Extension set up the program in New York, which resulted in 1.5 million gardens, aggregating a total of 200,000 acres.[13]

In 1948, the State University of New York was formed and the Colleges of Agriculture, Home Economics and Veterinary Medicine came into its decentralized structure. At that time the general supervision of these colleges was transferred from the Board of Regents to the Board of Trustees of the State University. In retrospect, there have been advantages and disadvantages of the change from time to time, but only minor effect upon the research and teaching of the colleges.[14]

---

12 Hedrick, Ulysses Prentiss *A History of Agriculture in the State of New York* p. 117-119
13 Bishop, Morris *A History of Cornell* p. 545-546
14 Colman, Gould P. *Education and Agriculture* p. 483-484

A view of the 1849 New York State Fair in Syracuse under the auspices of the New York State Agricultural Society. Courtesy of the Onondaga Historical Association.

In 1954, the relationship between Farm Bureau and the Extension Service was discontinued, ending an association that had existed for over 40 years. The County Farm and Home Bureau and 4-H Club Association became the County Extension Service Association. The Extension Service was funded by public funds making it inappropriate to encourage government legislation for the benefit of the farm sector. Farm Bureau, financed by farmers, now could assume that role and Extension continued its educational programs for farmers, homemakers and youth.

Cornell Cooperative Extension is the essence of cooperation between research and educators in bringing current research information to benefit the residents of New York. Cooperation also comes in funding for Extension, with a little over a third from county funds, a similar amount from the State and about an eighth from the federal government. The majority of its remaining income is from grants and contracts with local agencies. There are 55 county Extension Associations, each governed by a board of locally elected volunteers. Each of these Associations operate under a memoranda of agreement

with Cornell University to provide educational programming to residents of New York in areas such as agriculture, youth development, nutrition, environment and community development.

Extension offers programs and resources to benefit almost every aspect of agriculture including: production agriculture, agricultural economics and policy, environmental management, energy alternatives, food safety and agro-security, local foods, small farms, organic production, viticulture and enology.

The methods of bringing important information to New York farmers have changed dramatically since 1911 when John Barron visited farmers in his horse and buggy, but the importance of the information remains as timely and valuable as in 1911. Each year brings new challenges to the farmer, and Cornell Cooperative Extension is at his side helping to deal with them.

Cornell University has been blessed with outstanding presidents, deans, research scientists, Extension personnel and faculty. The New York State College of Agriculture has educated many great agricultural leaders, now located across the globe. Its research

Dairy cows being led down Tower Road at the New York State College of Agriculture at Cornell, circa 1910.

has solved innumerable problems and improved lives, not only for the residents of New York but for people throughout the world. Through Extension, the College has provided helpful information to millions of people. There is no question that Ezra Cornell's vision, of educating the common man has reaped benefits beyond his wildest dreams.

Although New York's agriculture is no longer the nation's leader, it has remained progressive and strong. The support of Cornell University, through its continued research, extension and teaching programs, and its progressive agricultural leaders and farmers, keep New York's agriculture on the cutting edge.

Although we automatically think of the Geneva Experiment Station as part of Cornell, it wasn't until 1923, when it was placed under the control of Cornell University. The Geneva Experiment Station's mission is to serve those who produce and consume New York's agricultural products. In its

early days it concentrated research on dairy, horticulture, and the evaluation of varieties of vegetable and field crops. A few years later it broadened its research to include beef cattle, swine and the evaluation of varieties of vegetable and fruit crops. Its evolvement continued with research activities in bacteriology, chemistry and plant diseases.

In 1923, when it was placed under the control of Cornell University, it expanded its research to include studies on canning crops, nursery plants and insect pests. At the end of World War II, all animal research was moved to Cornell and Geneva became a true horticultural research institute.

The Geneva Experiment Station places a strong emphasis on applied research but also maintains a balance with basic research and is recognized as having one of the most outstanding research programs in the world. In the 1980s, plant scientists at Geneva, in collaboration with researchers and engineers at the Cornell Nanofabrication facility, invented the "Gene Gun". This invention has enabled researchers to inject genetic material into the nuclei of plants, providing resistance to insects as well as resistance to salt and drought damage. Another example of their outstanding research was the development of the ultraviolet light pasteurization process, which is as effective and less costly than the traditional thermal pasteurization.

The Geneva facility encompasses over 700 acres of land devoted to test plots, orchards and vineyards, and has 20 major buildings. It employs 250 people doing research on 230 different projects. The Experiment Station has a budget of $16 million, approximately 60% funded by the State, with the remainder from the federal government, foundations, businesses, grower and food processor organizations and individuals.

In recent years the Food Technology Park was constructed at Geneva, which provides laboratory space to help businesses conduct research, while connecting with the resources available at Cornell. A number of businesses have used the facility and have produced a variety of items including an improved processed hay, all natural cookies, a healthy cherry drink, a deer-deterrent system and soil enhancement products.

The Cornell University Agricultural Experiment Station, located at Ithaca, links Cornell's world-class research facilities with one of the nation's most comprehensive statewide cooperative extension systems. Research ranges from childhood obesity to invasive species to global climate change. Cornell research dates back to the founding of the University and work by Isaac P. Roberts, Liberty Hyde Bailey and others. The Cornell Experiment Station manages 14,000 acres of farms and forests. It has more than a dozen farms, facilities and greenhouses providing critical research services to scientists. Today, its mission focuses on five areas: agriculture and food systems, quality of life, youth development, community and economic vitality and natural resources and the environment. It has progressed from research dealing with the needs of New York farm families, to the needs of people in all parts of the world.

Cornell's Experiment Station has a huge greenhouse complex, providing research on over 500 plant species. It conducts research on the majority of animals and plants, grown by New York's farmers, as well as many plants grown in other parts of the world. It even has a student-operated research farm using organic agricultural practices. Currently, there is a strong emphasis on developing renewable energy sources and improving the quality of New York's water resources.

The Cornell University Agricultural Experiment Station oversees a $7 million federally funded project portfolio in partnership with Cornell Cooperative Extension. The station provides research to benefit the instructional aspect of the college and also for the Extension Service, which brings the research knowledge to the end user in all parts of the State.

During the early 20th century there had been a movement to include agricultural, trade and industrial education in some of New York's high schools, and there was discussion whether this education should remain at that level or whether there should be an intermediary program at special institutes. Between the years 1907 and 1916, the latter choice was chosen with two-year agricultural and technical programs established at Canton, Alfred, Morrisville, Cobleskill, Delhi and Farmingdale.

In the book by William M. Houghton, *History and Development of Morrisville College,* Houghton noted that the high food prices of 1910 contributed to the establishment of the Agricultural and Technical Schools. Many of the hill farms in New York were being abandoned and the New York Legislature thought that improving the knowledge of the young men and women from farms would increase production and lower food costs. Other than for Farmingdale, on Long Island, the schools were located in areas where farms were being most rapidly abandoned.

In speaking of Morrisville, Houghton reiterates that the first class in the fall of 1910, was 50 students. To attract students it was important to keep the cost low. Room and board cost $4-5 a week with the total yearly cost, for the two 16-week terms of college classes, from $160-175. The College's policy was to provide work for many of the students and to pay them whatever they were worth, as long as it wasn't over 15 cents an hour! Morrisville and the other Ag & Tech schools filled an important educational need for rural youth and enrollments gradually grew. At the time these schools were formed, Liberty Hyde Bailey, Dean of the State College of Agriculture at Cornell, said that these colleges should train young men and women for farming, and Cornell should train the educators and other professionals. Bailey was appointed as a member of the administrative board at Morrisville, along with other Cornell representatives over the years, to help in the development of the new college.

Agriculture education is still offered at Morrisville and Cobleskill and has been extended to include bachelor degrees as well as the original associate degrees. Alfred offers degrees in agricultural business and agricultural technology. Delhi offers degrees in horticulture and turf management. Alfred and Delhi also offer courses in veterinary technology. Monroe Community College currently offers courses in agriculture and life sciences.

Traditionally, colleges of agriculture students were almost entirely from farms and rural areas. In the last few decades there has been a gradual change with more and more students from suburban and urban areas attending agricultural schools. There has been increased concern about the environment, food safety and conservation by the general population and colleges are offering programs that address these concerns. There has also been an increase of students, from other than rural areas, in production agriculture, especially organic, and production for green markets.

Another important area of agricultural education is offered in New York's high schools. The first vocational agricultural programs began to appear in the early 1900s and spread rapidly to rural schools throughout the State. The Future Farmers of America (FFA), normally formed from vocational education programs, was founded in 1920. About 300 schools offered vocational agriculture to 8,000 students in the early 1940s. During the World War II years, Congress passed a war food production-training program. 65,000 adults participated in this program offered by New York's agriculture vocational schools. Following the War, over 5,000 veterans received agricultural training. Today, the number of high school vocational education programs has decreased, but there are still over 90 FFA Chapters in New York, with some of the strongest chapters in suburban and urban schools.

Agricultural education is thriving and growing in New York. Although the number of people producing food has decreased dramatically over the past century, the increased awareness of the general population of the environment and its effect upon life is a compelling attraction. The number of student applications to colleges offering education in agriculture and life sciences, continues to increase. Some of our agricultural colleges still offer the traditional courses in production, marketing, business and technology, but all of them have broadened their offerings to include many non-traditional courses that appeal to young people in our rapidly changing world.

Chapter XVI
Governments' Role in Agriculture

Governments' role in agriculture was limited to local government in the late 18th century and well into the 19th century. Towns and villages often passed laws controlling animals allowed by owners to range on community lands or the property of other landowners. Laws often required animals to have a yoke on their neck to prevent them from going through a fence into a neighbor's crop and also limit the months of the year when animals could run free on public land. A system of ear notches, recorded in the local town, was used to identify animals. Laws, with substantial fines as penalties, required that, for example, Canada thistles be cut before going to seed.

As New York lands became increasingly populated and as the quantity of agricultural products increased, agricultural products were shipped greater distances to markets and laws were needed to protect both the producer and consumer who often were hundreds of miles apart. When products are shipped substantial distances there is greater competition in the marketplace as well as increased opportunity for deception and deterioration in quality. Laws and regulations were needed to protect individuals from deceptive practices and products that were not safe for consumption. Standards need to be developed to provide consistency of products with penalties established to prevent intentional deviations. For these reasons and many more both State and Federal governments have gradually increased their involvement with agriculture.

NEW YORK GOVERNMENT AND AGRICULTURE

Four and a half billion dollars of direct agricultural sales in New York provide the foundation of more than $25 billion toward the economy of the State. This tremendous economic input has caused numerous departments and state agencies to interact with New York's agricultural industry on a daily basis. A few examples are: Real Property Services that set equalization rates affecting the taxes paid on real estate; the Labor Department, which regulates farmers' relationships with employees; the Empire State Development Corporation that assists agricultural business working to increase the markets for New York's agricultural products; the New York State Energy Research and Development Authority that encourages energy conservation and the State Liquor Authority, which regulates wine products from over 200 New York wineries.

In *A History of American Agriculture: Government Programs and Policy*, it states that the Board of Agriculture, set up by the New York State Legislature, was the first organization of its kind in the United States. (The section on the New York State Agricultural Society in Chapter XIII provides more detail)

The department that interacts with agriculture in the most ways is the Department of Agriculture and Markets. A Department of Agriculture in New York was created in 1893 as successor to the appointment of Josiah Brown as Dairy Commissioner in 1884. The Department has had several name changes,

Plant Industry Exhibit by New York State Department of Agriculture and Markets at New York State Capitol in 1961. Photo Courtesy of New York State Archives.

becoming the Department of Farms and Markets in 1917 and in 1926 the Department of Agriculture and Markets. A governor appointee serves as the Commissioner of the department. This practice has usually resulted in a change of commissioners when a new Governor is elected. The Department has a dual role in serving both the farmer and the consumer and performs this role in many ways. The majority of the functions of the Department of Agriculture and Markets would come under four broad headings: economic development, environmental stewardship, food safety and security and consumer services. It would take an entire book to enumerate all of the functions of the department, so only a few of the key ways it interacts with agriculture and the consumer are mentioned here.

Landmark legislation was passed in 1971, at the urging of farmers, to provide for the formation of farmland protection boards, agricultural assessment of farmland, exemptions for farm buildings, protection of accepted agricultural practices and provision of a mechanism for a program to purchase development rights. It was recognized, by both the farmer and State government, that action was necessary to retain farmland for agricultural use. Through this program the Department oversees 289 Agriculture Districts in 53 counties, encompassing 22,823 farms covering 8.5 million acres.

When the law was enacted in 1971, an Agricultural District was required to be a minimum of 500 acres and all land in a district was required to be contiguous. Currently districts must contain a minimum of 500 acres, but the parcels no longer have to be contiguous. Non-farm owners can request their land be enrolled in an agricultural district and are eligible to receive an agricultural assessment if the land is leased by a farmer with gross farm sales over $10,000. Landowners in a district are permitted to request that their land be assessed based only on its agricultural value, without consideration for any development value. These agricultural assessments vary, depending upon the location of the land, from almost 100% of the land value in many rural areas to a very small percentage of the value in highly developed areas. These agricultural assessments have permitted farmland to remain agriculturally productive rather than to be sold for development because of the burden of taxes from high assessments. The law also provided for a penalty to be paid by the property owner if the farmland was converted to residential, commercial or industrial use. The landowner developing the farmland, and receiving an agricultural assessment, pays five times the taxes saved in the last year in which the land benefited from the assessment, plus six percent interest, compounded annually, for each year the agricultural assessment was granted, up to a maximum of five years.

Farmland protection boards have been established in many counties due to this legislation and are designed to identify specific strategies to maintain and enhance the viability of agriculture. The sale of development rights on thousands of acres, occurring over the last 13 years, is a direct result of this plan. In 2007, the State provided $35 million in funding, preserving 13,000 acres of farmland and in 2009, provided $23.1 million preserving 8,940 acres.

Another provision of the law is that governments give additional consideration in Agricultural Districts when taking land by eminent domain. The effect on agriculture is required to be minimal and the need for eminent domain must be critical.

An important responsibility of the Department, from an environmental standpoint, is the establishment of policies and the overseeing of programs for New York's 58 Soil and Water Conservation Districts. It also examines the potential environmental impact of gas and power lines as well as wind energy projects.

The Department also provides over 40,000 inspections of retail food establishments each year. Some of these inspections are farm, value-added enterprises where the farmer is providing an additional processing function such as turning some of his

Parading cattle on the race track of the New York State Fair, circa 1920. Courtesy of the New York State Fair.

Harold "Cap" Creal, Bernard Potter, both of whom served as New York State Fair directors, and Donald J. Wickham, Commissioner of New York State Ag & Mkts unveiling a plaque honoring "Cap" Creal at the New York State Fair, circa 1970. Photo courtesy of New York State Archives from New York State Ag & Mkts.

apples into pies or producing cheese from his milk. The inspections help insure that the consumer is receiving a safe food item and help the farmer by requiring that consistent standards be achieved. The majority of the inspections are of businesses not directly involved in farming but that do process food items.

A long-term function of the Department of Agriculture and Markets, along with the Industrial Exhibit Authority, is the operation of the New York State Fair. In 1900, the State accepted this responsibility from the New York State Agricultural Society, which had been operating the Fair since 1841. Each year there are approximately 40,000 contest entries, from 10,000 exhibitors, of animals and products for judging. Although the Fair lasts only 12 days, the fairgrounds are in constant use during the remainder of the year, for a variety of agricultural and public events. The Department also oversees the county fairs held throughout the State.

Advertising and promotion of the agricultural products produced by New York State farmers is another major function of the Department. It operates the Pride of New York program, which promotes New York-produced agricultural products, and has an organic program to help producers and processors take advantage of the demand for natural and organic food products. There is also a Grow New York program to aid in the development and expansion of agricultural businesses. Additionally, the Department works to help educate non-farm neighbors of active farms concerning the importance of agriculture and administers commodity-marketing orders that help growers with product development, research and promotion.

The Department provides protective services for both agriculture and the consumer in numerous ways. Both the Weights and Measures and the Grading and Certification divisions work to ensure purchasers receive what they expect. The Divisions of Plant and Animal Control work to protect our plants from potentially devastating diseases and invasive species. The Agriculture and Markets Law also requires the licensing of dealers who purchase more than $10,000 of goods a year from farmers.

In the 223 years since New York became a State, new laws affecting agriculture have been enacted on a yearly basis. Some of the laws have been beneficial, while others have placed added burdens upon agriculture. As the population of the State gradually changed from rural to urban and then to suburban and urban, it has become more difficult for the population to understand what is necessary to ensure New York State agriculture remains a healthy industry that can compete effectively with the agricultural production of other states, as well as other countries.

UNITED STATES GOVERNMENT AND NEW YORK AGRICULTURE

One of the early federal acts affecting agriculture, in an indirect manner, took place in 1790, with the establishment of a patent process for inventors. New York, soon to be the leading agricultural state, was blessed with many inventive minds that developed a multitude of patented labor-saving agricultural inventions.

The Embargo Act of 1807 is an early example of one of the many ways that the Federal government affects New York's agriculture. The Act forbade the departure of any merchant vessel from US ports. Since the Port of New York City had been the destination for great quantities of New York's agricultural products for shipment throughout the world, this Act created huge agricultural surpluses and disastrous prices for New York farm products.[1] Throughout the years, government action in Washington, over which the farmer has no control, has often made the difference between success and failure for generations of New York farmers.

A few of the hundreds of actions in Washington directly involving agriculture are: the Morrill Land Grant Act of 1862 that was critical in the establishment of Cornell University; a work projects for farm youth established in 1900, later becoming 4-H; meat inspection beginning in 1906; the Smith-Lever Act

in 1914, formalizing Cooperative Extension work; field mapping of soils beginning in 1899; development of commercial hybrid seed corn production in 1923; formation of the Farm Credit Administration in 1933 and establishing the first Soil Conservation District in 1937.

The United States Department of Agriculture (USDA) is the federal department that interacts with New York agriculture more than any other. Although a federal government agency, the USDA has had a tremendous effect upon the agriculture of New York. Established in 1862, during the presidency of Abraham Lincoln, the USDA affects the lives of all State residents on a daily basis. When it was established it lacked Cabinet status, which came in 1889. Initially, it was an agency designed to work with agriculture but today the bulk of its budget is designed for food and nutrition programs, like food stamps, which are for the benefit of any American, regardless of his or her employment. Its investment of many billions of dollars in research, over the years is one reason why we enjoy low-priced, high-quality food today.

The effects of USDA activities are in the food we eat and the environment in which we live. Almost all of the meat we eat, as well as many other foods, is processed under the supervision of the USDA. The quality of the water we drink and the water in our lakes and streams are enhanced by its activities.

There are two agencies within the USDA working with New York landowners and municipalities: the Farm Service Agency and the Natural Resources Conservation Service. Offices for both agencies and also Rural Development are at the same locations for the convenience of farmers and rural landowners. Rural Development is designed to increase economic opportunity and improve the quality of life for all rural Americans, rather than to be directly involved in agriculture.

The Farm Service Agency (FSA) administers and manages farm commodity, credit, conservation, disaster and loan programs as laid out by Congress through a network of federal, state and county offices. The programs are designed to improve the economic stability of the agricultural industry and help farmers adjust production to meet demand. It attempts to create a steady price range for agricultural commodities for both farmers and consumers.

1 Roberts, Elias H *American Commonwealths/New York* p. 505-506

The New York State office of the FSA is in Syracuse and there are 43 offices in New York State counties. Farm committee members, elected by farmers in the county districts, are responsible for equitably resolving local issues and are directly accountable to the Secretary of Agriculture. The county and state offices certify farmers for farm programs and pay out farm subsidies and disaster payments. The FSA also provides a Conservation Reserve Program with 10-to-15 year contracts, many of which buffer streams and wetlands in the State.

The FSA traces its beginnings to 1933, in the depths of the Great Depression, with the passage of the Agricultural Adjustment Act (AAA). The purpose of the AAA was to stabilize prices at a level at which farmers could survive. From this beginning there have been a variety of agencies formed, with a number of different names, in attempts to control the prices of agricultural products from being either excessively high or low. It is a never-ending battle as, for example, extreme weather conditions, in any large agricultural-producing area of the world, can affect prices.

Agriculture, throughout history, has had periods of dramatic price changes. The two World Wars, however, with the amazing mechanization of agriculture during the same time frame, created agricultural price fluctuations that required increased governmental intervention. During each of the World Wars, the government encouraged increased farm production and farmers responded with dramatic production increases, enhanced by the increased use of tractors, improved machinery, improved seeds and increased use of fertilizer. The "Genie" was out of the bottle and after each war, as a similar volume of production continued, there were huge agricultural surpluses with disastrously low prices. To manage these price fluctuations, a variety of programs were created to limit production, store surpluses, remove land from production and make payments to farmers to supplement market prices. The tripling of the world's population during the last 100 years has placed additional responsibility upon governments to continue to help ensure that agriculture continues to be capable of providing sufficient food.

The Natural Resources Conservation Service (NRCS), originally called the Soil Conservation Service, partners with landowners to conserve their soil, water and other natural resources. It provides engineering services for the installation of land tile, ponds and the lay out of contours. It works with landowners to protect the environment through conservation programs, which provide financial assistance to the landowner for environmental improvements. It also administers the Environmental and Quality Incentive Program, which provides funding toward the improvement of water quality through erosion and run-off control. Their goal, with the use of incentives, is to reward farms with the best practices and to motivate the others toward improved environmental stewardship. In addition, the NRCS has a wetland reserve program, which targets agricultural land that hasn't produced well, and pays the owner to turn it into wetlands and recreational use.

Although the Farm Credit System is independent and not operated by the US government, but since it was created by a series of legislative acts, it can be appropriately described in this section pertaining to government. The Farm Credit System dates back to 1908, when President Theodore Roosevelt appointed a Country Life Commission to address problems facing the rural population. At that time, agricultural real estate loans from commercial banks, if they were available at all, had prohibitively high rates and short terms. Until 1913, federal law prohibited national banks from making loans with maturities beyond five years. One of the most common ways for a farmer to borrow money for real estate was to go to a successful farmer in the community, and borrow the money from him, usually at 6% interest. This type of opportunity was often unavailable, however, because farmers with extra capital to loan were not readily available.

Debate over the appropriate legislation to help rural America was discussed in Congress for several years, with almost 100 pieces of legislation introduced, until 1916 when a cooperative structure was chosen. It was based upon 12 Federal Land Banks using $125 million in government seed money but financed by private capital from investors. Farm prices collapsed after World War I and the Federal Government added 12 Intermediate Credit Banks in an attempt to provide necessary credit to farmers, but these banks did not prove to be successful.

There were three pieces of legislation that followed, bringing sweeping reorganization to the Farm Credit

TABLE 34 – Average Value of New York Farms and Machinery & Equipment in Dollars
US Census

Year	Average Farm Value	Average Value Machinery & Equipment
1850	3,250	NA
1900	3,917	247
1950	11,742	NA
2007	449,010	97,550

NA Information not available

System. The first was the Agricultural Marketing Act of 1929, enacted to stabilize farm prices and also finance the development of farm cooperatives, the financing of which had been authorized earlier by the passage of the Capper-Volstead Act of 1922. The second was the Emergency Farm Mortgage Act, passed in 1933, which recapitalized the land banks with $189 million and cut interest rates to deal with the Depression. The third, also passed in 1933, was the Farm Credit Act, which established local Farm Credit Associations and 13 banks to service the thousands of cooperatives in the country. Also in 1933, President Franklin Roosevelt issued an executive order consolidating the supervision of all the federal credit agencies under the new Farm Credit Administration (FCA). These cooperatively owned financial entities, with the Farm Credit Administration as their regulator, form the basis of today's system.

Because of the continuing cycles of low and high agricultural prices the Farm Credit System has had challenging times. (Note the table on agricultural cycles in Chapter 7 concerning farm numbers, values and prices.) The System has had to borrow from the US Government in times of crisis but has repaid the money in full and continues to operate on a very sound basis today. There has been consolidation in the system, similar to both industry and agriculture, with only six Farm Credit System banks and a little more than 200 local lending associations remaining today. In 1991, Congress asked Farm Credit to play a greater role in financing agricultural marketing and processing operations, as well as water and sewer lines to rural communities. Strong rural communities are an important contributor to the prosperity of agriculture in those communities.

The information, in **Table 34**, taken from federal census information shows the changing need for credit for agriculture. It shows how drastically the average investment has increased for a farm and the machinery needed to equip it. It is not shown on this table, but large amounts of capital are needed annually to cover production costs, which averaged $96,372 in 2007. Note that the average value of farm equipment, on the average farm was only $247 in 1900.

The market value of products sold, by the average farm, in 2007, was $121,551. The fact that the farmer's investment in farm and machinery was over five times his gross annual income, prior to deducting production expenses of $96,372, underlines the need for the Farm Credit System. Farming is an extremely capital-intensive business.

The rapidly changing world of the 20th century, along with the even more rapidly changing world of the 21st century, require continual changes in both state and federal policies and programs to meet the needs of both agriculture and the consumer. It is a delicate balancing act and is always subject to government action that may be excessive, inappropriately delayed or ignored. Often the legislation and policies of the Federal government that are beneficial to other states are detrimental to New York or vice-versa. Government is agriculture's constant companion; at times its best friend and at other times its worst enemy.

Chapter XVII
New York's Agriculture in the 21ˢᵗ Century

Although New York's economy hasn't been dominated by agriculture for over a century, agriculture is still New York's number one industry and is extremely important in the life of our residents. Only a small portion of our population is aware of the complexity and size of our agricultural industry and the impact it has on their daily lives. The fact that New York agriculture is a $4½ billion industry doesn't give us the total dollars contributed to New York's economy. These dollars include purchases by farmers, additional jobs created across our entire economy, plus jobs created in the processing and the marketing of agricultural products. The capstone is the enhancement of our environment by the production of oxygen and the utilization of carbon dioxide from the atmosphere, by the thousands of acres of green crops grown on our farms. The open fields of growing crops and scattered woodlands provide vistas that artists attempt to replicate but can seldom achieve. These factors go far beyond man's ability to measure New York agriculture's value in dollars.

New York compares favorably, in its agriculture production, to the other 49 states. New York ranks third among the states in the production of milk and is first in the production of cottage cheese and sour cream. In fruit production New York ranks second in apples, third in grapes and fourth in tart cherries and pears. We are also one of the leading states in vegetable production, ranking second in cabbage, third in cauliflower, and fourth in pumpkins, snap beans and sweet corn. New York ranks second in the production of maple syrup and fifth in the production of Christmas trees. New York's diverse agricultural industry is a disadvantage, from a marketing standpoint, but an advantage to consumers who have a ready supply and range of fresh food products practically at their doorsteps. Although New York ranks 36ᵗʰ in the amount of farmland and is too often noted only for having the most populous city in the country, it has an agriculture industry of which we can be justifiably proud.

A segment of New York agriculture that has been growing, during the last decade, is organic agriculture. The International Federation of Organic Agriculture Movements (IFOAM) defines organic agriculture as follows: *organic agriculture is a production system that sustains the health of soils, ecosystems and people. It relies on ecological processes, biodiversity and cycles adapted to local conditions, rather than the use of inputs with adverse effects. Organic agriculture combines tradition, innovation and science to benefit the shared environment and promote fair relationships and a good quality of life for all involved.* The definitions of organic by the USDA and by State groups follow the guidelines of IFOAM, but go into greater detail and change with new information and need.

Organic agriculture was not a category in census information until the 2007 U.S. Census. The organic movement came about because of research showing the harmful effects of herbicides and insecticides to the endocrine systems of animals and humans. A great deal of research is ongoing and it is likely that organic agriculture will continue to grow until the effects of herbicides and insecticides are fully mitigated. Change is inevitable and perhaps with continued research, a "sustainable" standard will emerge that is a blend of conventional and organic practices. Organic sales nationally, as reported in the 2007 U.S. Census, were less than 1% of total agricultural sales. The dollar value of organic sales in the US totaled over 1.7 billion dollars with crops contributing two-thirds of that amount and livestock and poultry the other third.

New York's organic sales in the 2007 Census totaled $54 million with about 40% of that from crops and 60% from livestock and poultry. Of the $4.4 billion total agricultural sales in New York in 2007, organic sales were a little over 1%. Jefferson County had the greatest organic area with 7,500 acres, while Tompkins County had the greatest organic sales with over $5 million. Since New York has a large dairy industry, a large proportion of its organic production is in dairy products. Organic sales figures are not available since the 2007 Census information was obtained, but observations indicate that the movement toward greater consumption of organic food by the consumer is continuing.

Nationally, California is the organic leader with sales of organic agricultural products making up over one-third of national sales. Oregon is second, followed by Pennsylvania and New York. There were a total of 18,211 farms producing organic agricultural products with 1,137 in New York. Nationally there are 2 ½ million acres in organic production with 131,000 acres in New York.

It is ironic for me to see this movement toward organic since this was the manner in which almost all of the food in New York was produced into the 20th century. Up until the late 1940s, the author's farm had no herbicides and pesticides and the fertilizer was the organic waste from the cows, horses and turkeys. He was elated when corn could be sprayed to control weeds, eliminating the necessity of cultivating, hoeing and pulling weeds by hand. Both yield and labor efficiency increased with the use of these chemicals, providing a greater return for the farmer and lower prices for the consumer. Greater returns for the farmer were short-lived since when a method of increased production reaches across an industry, the increase in production lowers the price. The consumer, however, is ultimately the winner.

Organic food products normally cost more to produce and are priced at a premium to conventional food products. The production of organic products has often been of significant benefit to some small farms, by permitting continuation of the business at a profit, because of the premium price the farmer receives.

The production of organic farm products will probably continue to increase but will never approach the quantities produced by conventional methods because of the limited amount of natural fertilizer. The much smaller population of a century ago, along with millions of acres of fertile virgin soil, permitted the production of crops with little or no inorganic fertilizer. Today, without the use of inorganic fertilizer, millions of people would go hungry.

The movement toward organic production of food has brought increased awareness, to both the producer and the consumer, of the dangers of excessive use of chemicals in the production of our food. Farmers are becoming more careful to limit the amount of chemicals used, even when producing food using conventional methods.

The production of natural foods has also been increasing during the past two decades. The dictionary defines natural foods as foods that are minimally processed and contain no preservatives or artificial additives. Many farms have and continue to produce natural foods. Natural foods are sometimes organic, and sometimes they could be organic, but have not been certified by the appropriate authority. Soups, jams, preserves and other products are often produced as natural foods by small farms or home industries. These products are commonly found in green markets scattered throughout the State. There are also some large producers of natural food products that sell their products in large supermarkets and to specialty restaurants.

Natural food production, like organic production, is a niche market, filling a need for a minority of consumers and providing a livelihood for the relatively small producer. It is likely that there will be a continuing market for natural foods. A challenge for the natural food producer is the large manufacturers, who will see opportunity and provide competition that is difficult for the small producer to meet.

Another trend in recent years, has been the movement toward Community Supported Agriculture

(CSA). On CSA farms, consumers contract with the farmer for a weekly share of the crops produced. The produce is either picked up on the farm or delivered to the consumer. The consumer develops a closer relationship to the farm and is able to make firsthand observations of the farm's activities. There were 364 CSA farms in New York, reported in the 2007 U.S. Census, with Ulster County's 24 CSA farms first, followed by Suffolk and Columbia each with 17.

Many of our farms are not just producing crops but are adding value by providing additional processing. In 2007, there were 2,192 farms adding value to their production with further processing. An example of adding value is the farmer who produces milk and uses it to make yogurt for sale. Some other examples of adding value are the farmer that produces apples and uses some of them to make pies for sale; the chicken that is grown and processed ready for the consumer to cook, and the farms that produce soups ready to heat and eat, made from their own vegetables. These types of products, along with the movement toward organic and natural foods, have opened new markets for farmers desiring to add value to their products.

There has been an increase in the number of New York farms operated by families of the Amish and Mennonite faiths during the last 25 years. These farm families often sell some of their produce in retail outlets and also produce a variety of homemade foods that are sold directly to the consumer. Since these farms are not usually as highly mechanized as many of our other farms today, the farms tend to be smaller and produce a greater variety of crops. The number of these farms is likely to increase in the future.

Another recent trend in New York's agriculture is the generation of electricity on the farm. In 2007, there were 454 farms producing electricity, some of which was used on the farm and some sold to utilities, going directly into the electric grid. The most common method of producing electricity on the farm, at this time, is the production of methane from animal waste and then burning the methane to power generators. The remaining portion of the animal waste, after the production of methane, is used as fertilizer for crops. Experimental work is being done in growing hybrid willows and a variety of grasses for use in the production of ethanol from the plants' cellulose. If the research is successful, there is the potential for many additional New York acres to be utilized for the production of energy, similar to our 19th century production of energy, in the form of hay, for powering horses. A few farmers are also producing diesel fuel from the oil removed in crushing soybeans for the production of soybean meal. This, too, may be an area of agricultural growth by producing more energy for local consumption.

Income from dairying is New York's largest source of agricultural revenue with sales of over $2¼ billion. Dairying has been the state's largest agricultural enterprise for well over 100 years and will likely continue to be for some time in the future.

TABLE 35 – Comparison in the Production of New York Farm Products in 1900 and 2007

Product	1900	2007
Gallons of milk produced	773 million	1,407 million
Milk production per cow	4,426 pounds	19,303 pounds
Eggs produced	62 million dozen	93 million dozen
Apples produced	24 million bushels	31 million bushels
Grapes produced	124,000 tons	180,000 tons
Vegetable acres	107 thousand	160 thousand
Corn bushels	20 million	71 million*

* An additional 8.6 million tons of corn silage also produced in 2007

Note: 2007 information from New York Agricultural Statistics

Our climate and soils are favorable for the production of hay, pasture and corn silage, the major feeds for the dairy cow. The large number of consumers, within a few hours of our dairy farms, making milk delivery less costly than from the more distant states, is another reason that it is logical that dairying will continue as a major New York agricultural industry.

Although the number of New York farms is about one-sixth and the acreage in farms one-third of what existed 107 years ago, today our farms are producing more of several important crops. **Table 35** provides a comparison of the production of several farm items in 2007 compared to 1900.

The production per acre and production per animal have all increased dramatically, making the current production totals possible. **Table 35**, for example, only compares the production per cow for the two periods cited, but a production comparison is similar for all of our agricultural crops. Of the New York crops grown in 1900, the volume of production in 2007 has generally increased in the products where freshness is especially critical and the proximity of production to markets is important to consumers. Grain production, other than corn, has largely moved to the Mid-west, taking the production of livestock for meat consumption with it. New York's corn production continues to be important as a major feed for dairy cattle and poultry. The production of hops and tobacco has moved to areas of the country where conditions are more ideal than in New York. The market for oats and hay largely disappeared, in the early 1900s, with the advent of the automobile and truck that replaced thousands of horses.

New York's agriculture has always been diverse and though it is dominated by dairying, the varieties of production and the size of farms continue to increase. Today, we have family-owned and operated farms milking several thousand cows, growing vegetables on thousands of acres and producing eggs from more than 100,000 laying hens. In contrast, we also have farms with only a few cows, farms producing vegetables from an acre of land and farms with just a few laying hens producing specialty eggs. Each of these farms has found a niche that provides them with a market. We also have farms growing nursery crops that were unknown to New York 100 years ago, farms producing yogurt and farms creating new varieties of cheese. We have farms with alpacas and emus and farms producing tons of tomatoes and fresh lettuce even during the winter.

The percentage of the consumer's income spent for food has steadily decreased as farms have become increasingly more productive and efficient. The lower cost of food has released money for consumers to spend in other ways and on other things. It has made it possible for consumers to eat more often in restaurants, or when eating at home, to serve items that are fully prepared needing only to be taken out of a can or from the refrigerator and heated.

The increase of population in the State, from a few hundred thousand to 19 million, has consumed a large portion of our best land for homes, industry, highways and recreation. In the 1800s, land was cheap and there was lots of it, but as our population increases we find that land is not cheap and the quantity is limited. We are fortunate that there is land elsewhere and continually developing technology, enabling the production of sufficient food for our citizens. Nevertheless, we should give serious consideration as to whether it is appropriate to continue to remove fertile land, that cannot be replaced, from agricultural production.

New York's agriculture is a success story. Its soils have provided nourishment for many millions of people for hundreds of years. This has been accomplished through the hard work, sacrifices and dedication of over a dozen generations of farm families. Our soils are still rich and our farm families still rise to the task of providing bountiful harvests for the benefit of all. This is a legacy that brings pride and joy to our farming community and that will continue for generations to come. The residents of New York can take satisfaction in the agricultural industry of their state, which has continually served its citizens with a broad array of bountiful and healthy food. New York's agricultural industry is most certainly, "The Pride of New York".

Cutting ice on Otsego Lake, courtesy of New York History Museum.

Entrance to the Farmers' Museum at Cooperstown. This was originally a working dairy farm, with twin stone silos, constructed in about 1940 by the Clark family.

Wintertime mail delivery in the country, 1904. Courtesy of the OHA Museum & Research Center.

Neighbors having a plowing bee for a neighbor who was in ill health, circa 1913. It is customary practice for farmers to help each other in times of need. Courtesy of Jim Greene.

A horse powered sweep in operation, circa 1900. This sweep powered a machine to chop corn stalks. Courtesy of OHA Museum and Research Center.

Farmland under cultivation at Galen, in Wayne County, circa 1912. Courtesy of New York State Archives.

HAYING ON THE HUDSON

Haying on the Hudson showing a rope ferry, watercolor on paper, 1870. Crossing a river on a ferry, powered by a man pulling on a rope, was fairly common in the 19th century, when no bridge was available. Courtesy of Albany Institute of History and Art, 1968.47.98.

A farm in Rensselaer County, circa 1920. Courtesy of New York State Archives.

Threshing buckwheat in Schoharie County in 1911. Courtesy of New York State Archives.

Corn shocks and pumpkins in Otsego County in 1912. Courtesy of New York State Archives.

Filling a silo in St. Lawrence County in 1912. A steam engine, with water tank nearby, is furnishing power for the belt-driven ensilage cutter. Courtesy of New York State Archives.

Sharpening a scythe blade on a grindstone. One of the many jobs for a farm boy was turning the handle on the grindstone while his father sharpened tools. Because many farm jobs needed an extra hand, farm boys learned farming without ever realizing it. Courtesy of the Museum at the Shacksboro Schoolhouse, Baldwinsville, NY.

Ice wagon, delivering ice to village residents circa 1915. Ice was cut from lakes and ponds in the winter and stored in icehouses for delivery during the warm weather of summer. Courtesy of the Museum at the Shacksboro Schoolhouse, Baldwinsville, NY.

Appendix

TABLE A – Yards of Cloth Made by Families in New York Counties in 1810
in thousands

	Flax	Wool	Cotton
New York State	**5,373**	**3,258**	**216**
Suffolk	158	51	4
Queens	133	51	
Kings	32	4	
Richmond	23	2	
New York	<1	3	
Westchester	224	112	
Rockland	20	8	
Dutchess	230	129	
Columbia	342	255	21
Green	27	22	1
Ulster	222	87	7
Orange	212	96	4
Sullivan	35	14	
Delaware	131	71	1
Albany	145	87	
Rensselaer	222	158	2
Schenectady	5	3	
Saratoga	195	172	
Washington	351	384	51
Essex	34	70	
Clinton	24	24	
Franklin	10	5	
Montgomery	150	86	1
Schoharie	112	53	3
Herkimer	191	96	10
Oneida	13	160	81
Jefferson	107	51	1
St Lawrence	36	19	
Lewis	50	26	
Madison	225	120	5
Otsego	327	154	
Onondaga	196	107	3
Chenango	150	65	3
Cortland	39	25	3
Broome	63	23	
Tioga	67	25	2
Cayuga	217	120	
Seneca	116	49	2
Ontario	329	196	
Steuben	64	26	4
Allegany	12	3	<1
Genesee	89	30	2
Niagara	43	18	3

Data from US 1810 Census of American Manufactures

Note: The Census Marshall valued the flax at 37½ cents a yard totaling $2,014,742, the wool cloth at 87½ cents a yard totaling $2,850,585 and the cotton cloth at 32 cents a yard totaling $69,124.

TABLE B – Number of Tanneries and Hides Tanned in New York Counties in 1810

	Tanneries	Hides	Calf Skins	Sheep Skins
New York State	**867**	**151,165**	**210,445**	61,616
Suffolk	37	3,237	3,758	
Queens	10	4,005	7,412	
Kings	6	700	5,300	1,000
Richmond	2	2,600	1,100	
New York	9	10,300	17,500	43,400
Westchester	9	8,125	2,240	
Rockland	7	1,081	536	
Dutchess	80	14,240	28,474	
Columbia	34	5,334	11,666	
Green	25	6,433	6,516	
Ulster	41	4,123	8,429	1,574
Orange	49	10,780	11,348	1,690
Sullivan	11	596	1,671	158
Delaware	29	2,064	4,140	
Albany	31	5,220	12,548	4,307
Rensselaer	28	7,254	7,860	1,000
Schenectady				
Saratoga	33	8,962	3,478	1,324
Washington	37	10,000	14,000	
Essex	7	1,450	950	
Clinton	12	700	1,250	
Franklin		256	820	165
Montgomery	45	6,000	4,910	
Schoharie	14	2,739	1,576	
Herkimer	31	5,365	3,660	
Oneida	20	3,100	8,650	
Jefferson	16	750	1,000	
St Lawrence	12	600	1,167	
Lewis	11	800	1,250	
Madison	31	4,250	4,300	
Otsego	36	6,307	8,515	
Onondaga	31	1,855	7,076	
Chenango	8	996	662	
Cortland				
Broome	6	645	650	
Tioga	7	225	600	
Cayuga	19	1,900	3,630	
Seneca	15	744	3,058	
Ontario	37	6,600	7,000	
Steuben	5	173	745	
Allegany	2			
Genesee	7	656	1,000	
Niagara	17			

Data from US 1810 Census of American Manufactures

(blank spaces indicate no data available or no product produced)

Note: The Census Marshall placed a value on each of the hides at $4.25, the calfskins at $1.12 and the sheepskins at $1.15 giving a total value of $1,079,742.

TABLE C – Number of Naileries, Flaxseed Oil Mills & Breweries in New York Counties in 1810 and Quantities Produced

	Naileries		Flaxseed Oil Mills		Breweries	
	Number	**Pounds**	**Number**	**Gallons**	**Number**	**Gallons**
New York State	**44**	**2,292,960**	**28**	**33,427**	**42**	**2,004,504**
Suffolk						
Queens						
Kings						
Richmond						
New York	4	139,440			15	1,528,872
Westchester	1	38,080	2	NA		
Rockland	1	1,002,400				
Dutchess			3	3,500	2	18,000
Columbia					1	NA
Green	4	31,680				
Ulster	7	NA				
Orange	2	120,960			1	2,000
Sullivan						
Delaware	2	8,960				
Albany	2	324,800			5	297,000
Rensselaer	5	232,960	3	6,287	1	25,600
Schenectady						
Saratoga	1	5,600			1	12,800
Washington					1	4,000
Essex	2	6,720				
Clinton	2	4,480				
Franklin						
Montgomery	3	7,840				
Schoharie	1		3	1,800		
Herkimer	1	1,120	1	NA		
Oneida	3	224,000	1	700	4	76,800
Jefferson			3	9,650	2	25,600
St Lawrence						
Lewis						
Madison						
Otsego	6	100,800	2	640	1	NA
Onondaga	2	38,080	3	3,750	2	7,232
Chenango	1	3,360	1	300	4	7,000
Cortland			1	300		
Broome						
Tioga	1	1,680			3	4,500
Cayuga						
Seneca						
Ontario						
Steuben						
Allegany						
Genesee					1	1,600
Niagara						

Data from 1810 US Census of American Manufactures

Note: Naileries were often blacksmith shops where nails were made one at a time.
The Census Marshall put a value on the nails of twelve cents a pound totaling $275,155, the flaxseed oil at $1.25 a gallon totaling $41,783 and the 2,004,504 gallons from the breweries at seventeen cents a gallon totaling $340,767.

TABLE D – Population of New York State Counties Using Boundaries as They Existed in 1865

(in thousands) From recapitulation by counties in introduction of 1865 NY Census

Year	1790	1800	1810	1820	1830
New York State	340	588	961	1,302	1,913
Albany	13	25	35	38	53
Allegany			1	6	20
Broome		2*	6	11	17
Cattaraugus			<1	4	16
Cayuga		10	29	38	47
Chautauqua			2	12	34
Chemung	2*	1*	2*	4*	11*
Chenango		6	21	31	37
Clinton	1	3	8	12	19
Columbia	27	35	32	38	39
Cortland		<1*	8	16	23
Delaware	2*	10*	20	26	33
Dutchess	36	37	41	46	50
Erie			4*	10*	35
Essex	<1*	4	9	12	19
Franklin		<1*	2	4	11
Fulton		6*	15*	15*	20*
Genesee			3	18	26
Greene	7*	12	19	22	29
Hamilton			<1	1	1
Herkimer	2*	16	24	31	35
Jefferson			15	32	48
Kings	4	5	8	11	20
Lewis		1	6	29	15
Livingston		2*	10*	31*	27
Madison		8*	25	32	39
Monroe		1*	4*	27*	49
Montgomery	18	13	23	23	23
Nassau (a)					
New York	33	60	96	121	197
Niagara			1	7	18
Oneida	1*	20	30	50	71
Onondaga		6	25	41	58
Ontario	1	8	22	35	40
Orange	18	29	34	41	45
Orleans			1*	5	17
Oswego			3*	12	27
Otsego	1*	21	38	44	51
Putnam	8*	9*	10*	11	12
Queens	16	16	19	21	22
Rensselaer	22*	30	36	40	49
Richmond	3	4	5	6	7
Rockland	6*	6	7	8	9
St. Lawrence			7	16	36
Saratoga	17*	24	33	36	38
Schenectady	5*	8*	10	13	12
Schoharie	2*	9	18	23	27
Schuyler			3*	10*	13*
Seneca		4*	11	17	21
Steuben		1	6	18	28
Suffolk	16	19	21	24	26
Sullivan	1*	3*	6	8	12
Tioga		2	5	7	13
Tompkins			5*	22	31
Ulster	16	21	26	30	36
Warren	1*	4*	7*	9	11
Washington	20	30	36	38	42
Wayne		1*	6*	20*	33
Westchester	24	27	30	32	36
Wyoming			2*	16*	29*
Yates		1*	4*	11*	19

TABLE E – Population of New York Counties
(in thousands) US Census

Year	1840	1850	1900	1950	2000
New York State	2,428	3,097	7,268	14,830	18,976
Albany	68	93	165	239	294
Allegany	30	37	41	43	49
Bronx				1,451	1,332
Broome	22	30	69	184	200
Cattaraugus	28	38	65	77	83
Cayuga	50	55	66	70	81
Chautauqua	47	50	88	135	139
Chemung	15	21	54	86	91
Chenango	40	40	36	39	51
Clinton	28	40	47	53	79
Columbia	43	43	43	43	63
Cortland	24	25	27	37	48
Delaware	35	39	46	44	48
Dutchess	52	58	81	136	280
Erie	62	100	433	899	950
Essex	23	31	30	35	38
Franklin	16	25	42	44	51
Fulton	18	20	42	51	55
Genesee	28	28	34	47	60
Greene	30	33	31	28	48
Hamilton	1	2	4	4	5
Herkimer	37	38	51	61	64
Jefferson	60	68	76	85	111
Kings	47	138	1,166	2,738	2,465
Lewis	17	24	27	22	26
Livingston	42	40	37	40	64
Madison	40	43	40	46	69
Monroe	64	87	217	487	735
Montgomery	35	31	47	59	49
Nassau			55	672	1,334
New York	312	515	2,050	1,960	1,537
Niagara	31	42	74	189	219
Oneida	85	99	132	222	235
Onondaga	67	85	168	341	458
Ontario	43	43	49	60	100
Orange	50	57	103	152	341
Orleans	25	28	36	29	44
Oswego	43	62	70	77	122
Otsego	49	48	48	50	61
Putnam	12	14	36	20	95
Queens	30	36	152	1,550	2,229
Rensselaer	60	73	121	132	152
Richmond	10	15	67	191	443
Rockland	11	16	38	89	286
St. Lawrence	56	68	89	98	111
Saratoga	40	45	61	74	200
Schenectady	17	20	46	142	146
Schoharie	32	33	26	22	31
Schuyler	16*	18*	15	14	19
Seneca	24	25	26	29	33
Steuben	40	58	82	91	98
Suffolk	32	36	77	276	1,419
Sullivan	15	25	32	40	73
Tioga	20	24	27	30	51
Tompkins	32	32	33	59	96
Ulster	45	59	88	92	177
Warren	13	17	29	39	63
Washington	41	44	45	47	61
Wayne	42	44	48	57	93
Westchester	48	58	184	625	923
Wyoming	31*	31	30	32	43
Yates	20	20	20	17	24

(*) Not formed at this time

TABLE F – Population of the States
(in thousands) US Census

Year	1790	1800	1810	1820	1830	1840
New York	340	586	959	1,372	1,918	2,428
Maine	96	151	228	298	399	501
New Hampshire	141	183	214	244	269	284
Massachusetts	378	423	472	523	610	737
Rhode Island	69	69	77	83	97	108
Connecticut	238	251	262	275	297	309
Vermont	85	154	217	235	280	291
New Jersey	184	211	245	277	320	373
Pennsylvania	431	602	810	1,049	1,348	1,724
Delaware	59	64	72	72	76	78
Maryland	319	341	380	407	447	470
Virginia	748	880	974	1,065	1,211	1,239
North Carolina	393	478	555	638	737	753
South Carolina	249	315	415	502	581	594
Georgia	82	162	252	340	516	691
Kentucky	73	220	406	561	687	779
Tennessee	35	105	261	422	621	829
Ohio		45	230	581	937	1,519
Indiana		4	24	147	343	685
Mississippi		8	40	75	136	375
District of Columbia		14	24	33	39	43
Illinois			12	55	157	476
Michigan			4	8	31	212
Louisiana			76	153	215	352
Missouri			20	66	140	383
Alabama				127	309	590
Florida				14	34	54
Wisconsin						30
Iowa						43
United States	**3,929**	**5,305**	**7,239**	**9,638**	**12,866**	**17,069**

TABLE G – Production by State
(in thousands) 1840 Federal Census

State	Cords of Wood Sold	Dollars of Dairy Products	Dollars of Orchard Products	Dollars of Family Goods	Dollars of Market Produce	Dollars of Nursery Produce
New York	1,059	10,496	1,702	4,637	499	76
Maine	205	1,497	149	804	52	<1
New Hampshire	116	1,638	240	538	18	<1
Massachusetts	278	2,373	389	232	284	112
Rhode Island	49	223	32	51	68	13
Connecticut	159	1,377	296	226	62	18
Vermont	96	2,009	214	675	16	6
New Jersey	341	1,328	464	202	250	26
Pennsylvania	270	3,187	618	1,303	233	50
Delaware	68	114	28	62	4	1
Maryland	178	457	106	176	133	11
Virginia	404	1,480	706	2,442	92	39
North Carolina	40	674	386	1,413	28	49
South Carolina	171	578	52	931	38	2
Georgia	57	605	156	1,468	19	2
Alabama	61	265	55	1,656	32	<1
Mississippi	118	360	14	683	43	<1
Louisiana	203	153	12	65	240	32
Tennessee	104	472	367	2,887	20	71
Kentucky	264	931	435	2,622	125	6
Ohio	273	1,849	475	1,854	98	20
Indiana	184	742	110	1,290	61	17
Illinois	135	428	127	994	72	23
Missouri	82	100	91	1,150	37	6
Arkansas	79	59	11	490	3	<1
Michigan	54	301	16	114	4	6
Florida	10	23	1	20	12	<1
Wisconsin	23	36	<1	13	3	1
Iowa	7	24	<1	26	2	4
District of Columbia	1	6	4	2	53	<1
US	**5,089**	**33,787**	**7,257**	**29,023**	**2,601**	**593**

TABLE H – Products by State
(in thousands) 1840 Federal Census

State	Wheat in bushels	Barley in bushels	Oats in bushels	Corn in bushels	Potatoes in bushels	Hops in pounds
New York	12,286	2,520	20,676	10,972	30,124	447
Maine	848	355	1,076	951	10,392	37
New Hampshire	422	122	1,296	1,163	6,207	243
Massachusetts	158	165	1,319	1,809	5,386	255
Rhode Island	3	66	172	450	912	<1
Connecticut	87	34	1,453	1,500	3,414	5
Vermont	496	55	2,223	1,120	8,870	48
New Jersey	774	13	3,084	4,362	2,072	5
Pennsylvania	13,213	210	20,642	14,240	9,536	49
Delaware	315	5	927	2,099	201	<1
Maryland	3,346	4	3,534	8,233	1,036	2
Virginia	10,110	87	13,451	34,578	2,945	11
North Carolina	1,961	4	3,194	23,894	2,609	1
South Carolina	968	4	1,487	14,723	2,698	<1
Georgia	1,802	13	1,610	20,905	1,291	<1
Alabama	828	8	1,406	20,947	1,708	<1
Mississippi	197	2	669	13,161	1,630	<1
Louisiana	<1	0	107	5,953	834	<1
Tennessee	4,570	5	7,036	44,986	1,904	<1
Kentucky	4,803	17	7,156	39,847	1,055	<1
Ohio	16,572	212	14,393	33,668	5,805	62
Indiana	4,049	28	5,982	28,156	1,526	39
Illinois	3,335	82	4,988	22,634	2,026	18
Missouri	1,037	10	2,234	17,332	784	<1
Arkansas	106	<1	189	4,847	294	0
Michigan	2,157	128	2,114	2,277	2,109	11
Florida	<1	<1	14	899	265	0
Wisconsin	212	11	407	379	420	<1
Iowa	155	<1	216	1,406	234	<1
District of Columbia	12	<1	16	39	12	<1
US	**84,823**	**4,162**	**123,071**	**377,532**	**108,298**	**1,239**

TABLE I – Products by State
(in thousands) 1840 Federal Census

State	Horses & Mules	Neat Cattle	Sheep	Swine	Est. Value of Poultry	Tons of Hay
New York	475	1,911	5,119	1,900	1,153	3,127
Maine	59	327	649	117	123	691
New Hampshire	44	276	617	122	107	496
Massachusetts	61	283	378	143	178	569
Rhode Island	8	37	90	31	62	63
Connecticut	35	239	403	132	177	427
Vermont	62	384	1,682	204	132	837
New Jersey	71	220	219	261	337	335
Pennsylvania	365	1,173	1,768	1,504	686	1,312
Delaware	14	54	39	74	47	22
Maryland	92	226	258	417	219	107
Virginia	326	1,024	1,294	1,992	755	365
North Carolina	167	617	538	1,650	544	101
South Carolina	130	573	233	879	396	25
Georgia	158	884	267	1,458	450	17
Alabama	143	668	163	1,424	405	13
Mississippi	109	623	128	1,001	369	<1
Louisiana	100	381	98	323	284	25
Tennessee	341	823	742	2,927	607	31
Kentucky	396	787	1,008	2,311	536	88
Ohio	431	1,218	2,028	2,100	551	1,022
Indiana	241	620	676	1,624	358	178
Illinois	199	626	396	1,495	309	165
Missouri	196	434	348	1,271	271	49
Arkansas	51	189	42	393	109	<1
Michigan	30	185	100	296	83	131
Florida	12	118	7	93	61	1
Wisconsin	6	30	3	51	16	31
Iowa	11	38	15	105	17	18
District of Columbia	2	3	<1	5	3	1
US	**4,336**	**14,972**	**19,311**	**26,301**	**9,344**	**10,248**

TABLE J – Acres of Improved Farm Land by County
(in thousands) NY Census

Year	1825	1835	1845	1855	1865	1875
New York State	7,160	9,655	11,757	13,657	14,827	15,875
Albany	191	207	233	242	267	248
Allegany	61	139	204	280	310	374
Broome	60	93	144	198	230	271
Cattaraugus	22	87	157	266	323	380
Cayuga	187	257	295	315	323	336
Chautauqua	67	167	252	360	380	432
Chemung			104	120	133	146
Chenango	178	248	309	347	378	396
Clinton	62	103	125	168	209	237
Columbia	264	307	311	304	310	315
Cortland	84	118	160	194	207	218
Delaware	157	224	307	364	427	452
Dutchess	368	365	379	366	361	373
Erie	75	162	224	340	407	458
Essex	77	113	206	185*	246	204
Franklin	30	59	121	144	183	214
Fulton			119	133	133	140
Genesee	170	305	194	219	226	238
Greene	142	174	199	212	221	235
Hamilton			11	16	20	24
Herkimer	183	226	255	267	278	297
Jefferson	173	258	386	465	514	556
Kings	24	22	20	15	14	9
Lewis	47	86	114	184	211	241
Livingston	113	177	214	262	281	295
Madison	176	223	267	277	285	301
Monroe	136	229	281	216*	324	332
Montgomery	243	305	190	194	196	200
New York	5	4	4	1	>1	
Niagara	42	98	148	207	224	257
Oneida	224	300	362	435	460	501
Onondaga	193	270	311	344	343	373
Ontario	183	231	274	290	318	314
Orange	243	294	302	308	303	321
Orleans	51	117	151	181	186	192
Oswego	52	110	166	244	274	313
Otsego	268	341	389	428	435	454
Putnam	79	90	104	94	87	87
Queens	144	129	125	119	119	117
Rensselaer	234	263	278	292	290	299
Richmond	22	21	17	15	12	11
Rockland	45	62	55	46	49	43
St. Lawrence	104	151	305	499	602	712
Saratoga	253	288	295	315	324	317
Schenectady	67	78	92	93	94	95
Schoharie	143	183	234	227	253	280
Schuyler				134	139	147
Seneca	94	131	140	151	162	165
Steuben	112	183	277	361	426	503
Suffolk	167	168	157	163	148	156
Sullivan	41	50	68	125	156	191
Tioga	83	139	103	154	176	199
Tompkins	135	186	223	205	206	223
Ulster	160	185	216	240	253	272
Warren	52	71	83	111	155	136
Washington	252	308	310	333	338	346
Wayne	91	153	206	254	259	284
Westchester	239	255	230	209	193	174
Wyoming			180	241	263	281
Yates	66	115	140	155	151	163

* Improved farmland decreased for no apparent reason other than possible error or change in definition   **Note:** Chemung formed from Tioga in 1836; Fulton from Montgomery in 1838; Schuyler from Chemung, Steuben & Tompkins in 1854; Wyoming from Genesee in 1841

TABLE K – Number of New York Farms by County
US Census

Year	1860	1900	1950	1997	2007
New York State	195,459	226,720	124,977	31,757	36,352
Albany	3,194	3,281	1,453	396	498
Allegany	4,448	5,082	2,604	724	847
Broome	2,786	4,410	2,345	511	580
Cattaraugus	4,500	5,976	3,508	946	1,122
Cayuga	4,455	5,039	2,934	846	936
Chautauqua	5,890	7,404	5,336	1,557	1,658
Chemung	1,599	2,438	1,280	313	373
Chenango	4,637	4,473	2,689	801	908
Clinton	2,977	3,764	2,275	488	590
Columbia	2,920	2,944	1,692	464	554
Cortland	2,797	2,754	1,414	452	587
Delaware	4,965	5,232	3,234	717	747
Dutchess	3,356	3,537	1,729	539	656
Erie	6,219	7,929	4,611	973	1,215
Essex	2,228	2,412	1,156	197	243
Franklin	3,101	3,721	2,074	476	604
Fulton	1,609	2,234	830	176	222
Genesee	2,845	3,286	2,112	516	551
Greene	2,535	2,746	1,300	244	386
Hamilton	308	510	75	13	20
Herkimer	3,078	3,227	1,869	583	672
Jefferson	6,097	6,052	3,440	916	885
Kings	517	360	65	8	1
Lewis	2,596	3,838	1,701	623	616
Livingston	2,829	3,267	1,835	625	792
Madison	3,858	4,144	2,360	692	744
Monroe	4,624	5,889	3,147	480	585
Montgomery	1,994	2,407	1,473	542	604
Nassau	#	1,658	618	55	59
New York	118	184		2	
Niagara	3,257	4,356	3,362	687	865
Oneida	6,768	7,232	3,909	928	1,013
Onondaga	5,169	6,305	3,405	602	692
Ontario	3,363	4,328	2,507	692	859
Orange	3,176	3,966	2,958	624	642
Orleans	2,406	2,964	1,878	456	554
Oswego	4,995	6,914	3,339	605	639
Otsego	5,695	5,634	3,261	865	980
Putnam	1,570	1,141	319	48	72
Queens	2,417#	1,188	111	2	4
Rensselaer	2,999	3,668	1,822	459	506
Richmond	216	290	113	7	14
Rockland	764	939	408	21	21
St. Lawrence	8,002	8,353	5,091	1,363	1,330
Saratoga	3,769	3,805	1,795	472	641
Schenectady	1,203	1,194	600	151	194
Schoharie	3,356	3,437	1,940	518	525
Schuyler	2,322	2,103	1,118	318	394
Seneca	1,935	2,303	1,275	413	513
Steuben	5,903	8,179	3,833	1,295	1,578
Suffolk	3,022	3,277	2,187	606	585
Sullivan	2,978	3,887	1,881	311	323
Tioga	2,694	3,134	1,870	497	565
Tompkins	2,918	3,270	1,662	447	588
Ulster	3,879	5,184	2,552	409	501
Warren	1,772	2,121	547	58	86
Washington	3,471	3,715	2,349	738	843
Wayne	4,044	5,286	3,643	840	938
Westchester	3,252	2,326	664	91	106
Wyoming	3,428	3,519	2,217	702	761
Yates	1,940	2,504	1,183	657	864

(D) Data withheld to protect privacy   #Nassau County was formed from Queens in 1899

196

TABLE L – Acres and Values
1850 US Census

States	Farm Acres in (000s)		Value in ($000s)			
	Improved	Unimproved	Farms	Machinery	Livestock	Animals slaughtered
New York	12,408	6,710	554,546	22,084	73,570	13,573
Maine	2,039	2,515	54,861	2,284	9,705	1,646
New Hampshire	2,251	1,140	55,245	2,314	8,871	1,522
Vermont	2,601	1,524	63,367	2,739	12,643	1,861
Massachusetts	2,133	1,222	109,076	3,209	9,647	2,500
Rhode Island	356	197	17,070	497	1,532	667
Connecticut	1,768	615	79,726	1,829	7,467	2,202
New Jersey	1,767	984	120,237	4,425	10,679	2,638
Pennsylvania	8,698	6,294	407,876	14,722	41,500	8,219
Delaware	580	375	18,880	510	1,849	373
Maryland	2,797	1,836	87,178	2,463	7,997	1,954
Dist. Columbia	16	11	1,730	40	71	9
Virginia	10,360	15,792	216,401	7,021	33,656	7,502
North Carolina	5,543	15,543	67,891	3,931	17,717	5,767
South Carolina	4,079	12,145	82,431	4,136	15,060	3,502
Georgia	6,378	16,442	95,753	5,894	25,708	6,339
Florida	349	1,246	6,393	658	2,880	514
Alabama	4,435	7,702	64,323	5,125	21,690	4,823
Mississippi	3,444	7,046	54,738	5,762	19,403	3,636
Louisiana	1,590	3,399	75,814	11,576	11,152	1,458
Texas	643	10,852	16,550	2,151	10,412	1,116
Arkansas	781	1,816	15,265	1,601	6,647	1,163
Tennessee	5,175	13,608	97,851	5,360	29,978	6,401
Kentucky	5,968	10,981	155,021	5,169	22,661	6,462
Missouri	2,938	6,794	63,225	3,981	19,887	3,367
Illinois	5,039	6,997	96,133	6,405	24,209	4,972
Indiana	5,046	7,746	136,385	6,704	22,478	6,567
Ohio	9,851	8,146	358,758	12,750	44,121	7,439
Michigan	1,929	2,454	51,872	2,891	8,008	1,328
Wisconsin	1,045	1,931	28,528	1,641	4,897	920
Iowa	824	1,911	16,657	1,172	3,689	821
California	32	3,861	3,874	103	3,351	107
Minnesota*	5	23	161	15	92	2
New Mexico*	166	124	1,653	77	1,494	82
Oregon*	132	299	2,849	183	1,876	164
Utah*	16	30	311	84	546	67
Total	**113,032**	**180,528**	**3,271,575**	**151,587**	**544,180**	**111,703**

* Territories not yet formed into states

TABLE M – Average Acreage of New York Farms by County
US Census

Year	1860	1900	1950	1997	2007
New York State	107	100	128	228	197
Albany	95	91	111	143	123
Allegany	114	116	160	218	178
Broome	111	95	126	168	149
Cattaraugus	123	113	146	203	163
Cayuga	91	82	121	298	267
Chautauqua	102	83	94	157	142
Chemung	121	90	122	189	175
Chenango	114	122	155	229	195
Clinton	106	116	176	305	253
Columbia	137	128	154	248	192
Cortland	106	110	169	267	213
Delaware	147	152	195	256	222
Dutchess	141	132	176	198	156
Erie	85	72	82	147	123
Essex	199	167	170	245	207
Franklin	104	115	147	342	217
Fulton	124	93	111	195	152
Genesee	99	90	122	331	333
Greene	131	123	132	200	155
Hamilton	167	125	97	61	23
Herkimer	131	119	151	243	208
Jefferson	132	123	170	318	296
Kings	33	18	3	1	(D)
Lewis	110	129	189	288	272
Livingston	108	114	178	316	281
Madison	93	94	135	269	253
Monroe	81	65	91	215	227
Montgomery	118	98	139	249	206
Nassau	#	53	44	25	22
New York	13	19		(D)	
Niagara	93	70	74	185	165
Oneida	93	91	124	233	190
Onondaga	88	72	98	244	217
Ontario	113	94	121	269	232
Orange	120	101	93	152	126
Orleans	95	80	107	314	252
Oswego	79	71	101	169	157
Otsego	109	109	147	239	180
Putnam	112	117	135	72	78
Queens	66	22	5	(D)	(D)
Rensselaer	116	100	125	216	168
Richmond	68	40	18	4	(D)
Rockland	87	66	43	27	(D)
St. Lawrence	106	128	174	291	261
Saratoga	112	107	112	155	118
Schenectady	99	100	101	120	99
Schoharie	110	107	149	214	182
Schuyler	87	94	128	205	168
Seneca	95	84	121	284	249
Steuben	121	101	165	269	236
Suffolk	121	84	56	59	59
Sullivan	109	123	102	187	156
Tioga	100	97	132	220	189
Tompkins	92	87	122	214	185
Ulster	112	101	89	169	150
Warren	137	135	135	158	99
Washington	128	122	164	264	241
Wayne	60	69	87	199	180
Westchester	80	79	73	83	80
Wyoming	101	105	147	278	287
Yates	105	81	132	159	146

(D) Data withheld to protect privacy   # Nassau County was formed from Queens in 1899

TABLE N – Milk Production by County in 1899 and in 2006
1899 figures from US Census and 2006 figures from NY Ag & Mkts. Estimates

County	Millions of gallons 1899	Millions of gallons 2006
Albany	8	3
Allegany	18	17
Broome	16	16
Cattaraugus	29	37
Cayuga	14	84
Chautauqua	25	42
Chemung	7	6
Chenango	29	30
Clinton	9	41
Columbia	9	14
Cortland	15	29
Delaware	43	20
Dutchess	18	5
Erie	24	30
Essex	5	3
Franklin	12	33
Fulton	5	6
Genesee	7	56
Greene	9	2
Hamilton	1	NA
Herkimer	20	26
Jefferson	32	71
Kings	1	NA
Lewis	17	57
Livingston	6	47
Madison	19	36
Monroe	11	9
Montgomery	10	42
Nassau	2	NA
New York	1	NA
Niagara	8	22
Oneida	31	48
Onondaga	18	47
Ontario	8	45
Orange	32	12
Orleans	4	6
Oswego	18	5
Otsego	26	30
Putnam	6	NA
Queens	4	NA
Rensselaer	11	25
Richmond	1	NA
Rockland	1	NA
St Lawrence	49	76
Saratoga	8	19
Schenectady	3	1
Schoharie	13	13
Schuyler	3	10
Seneca	4	13
Steuben	15	41
Suffolk	4	NA
Sullivan	9	6
Tioga	11	16
Tompkins	8	20
Ulster	10	NA
Warren	3	NA
Washington	11	43
Wayne	10	18
Westchester	8	NA
Wyoming	12	121
Yates	3	24
New York total	**773**	**1,424**

NA Not available (Combined with other counties or no milk production)

TABLE O – Comparisons of New York with the US and other States in Various Years
US Census
Population, number of farms, and total farm acres are in thousands (000)

		1850	1900	1950	2007
Population	US	23,192	76,212	151,326	301,621
Population	NY	3,097	7,269	14,830	19,297
Percent	NY	13.4	9.5	9.8	6.4
Rank	NY	1	1	1	3
Number of Farms	US	1,449	5,790	5,382	2,205
Number of Farms	NY	171	227	125	34
Percent of Farms	NY	11.8	3.9	2.3	1.5
Rank among States	NY	1	7	23	27
Total Farm Acres	US	293,561	841,262	1,158,566	922,096
Total Farm Acres	NY	19,119	22,648	16,017	7,175
Percent of Acres	NY	6.5	2.7	1.4	0.8
Rank among States	NY	4	12	30	36

TABLE P – Comparisons of New York with the US and other States in Various Years
US Census Information
Acres are all in thousands (000)

		1850	1900	1950	2007
Number of Milk Cows	US	6,385	17,190	21,232	9,267
Number of Milk Cows	NY	931	1,502	1,218	626
Percent of Milk Cows	NY	14.5	8.7	5.7	6.8
Rank among States	NY	1	1	3	3
Wheat Acres	US	100,486 (a)	52,589	50,826	50,933
Wheat Acres	NY	13,121 (a)	558	385	85
Percent of Acres	NY	13.1 (a)	1.1	0.7	0.02
Rank among States	NY	3 (a)	20	18	35
Corn Acres	US	592,071 (a)	94,913	83,336	92,228
Corn Acres	NY	17,858 (a)	659	624	1,059
Percent of Acres	NY	3.0 (a)	0.7	0.7	1.1
Rank among States	NY	13 (a)	27	26	19
Potato Acres	US	108,298 (b)	2,939	1,514	1,132
Potato Acres	NY	30,124 (b)	396	112	19
Percent of Acres	NY	27.8 (b)	13.5	7.4	1.7
Rank among States	NY	1 (b)	1	3	13

(a) Bushels instead of acres   (b) 1840 US Census and bushels instead of acres

TABLE Q – Changes in New York Agriculture
US Census

Year	Number of Farms	Land in Farms (000's acres)	Value of Farms (000's $)	Total Farm Sales (000's $)	Livestock Sales (000's $)	Crop Sales (000's $)
1850	170,621	19,119	554,547			
1860	196,990	20,975	803,344			
1870	216,253	22,191	1,018,286			
1880	241,058	23,781	1,056,177			
1890	226,223	21,962	968,127			
1900	226,720	22,648	888,134			
1910	215,597	22,030	1,186,746			
1920	193,195	20,633	1,425,062			
1930	159,806	17,980	1,315,905	342,544	246,244	96,300
1940	153,238	17,170	947,074	242,417	167,052	75,365
1950	124,977	16,016	1,467,452	630,400	466,248	164,152
1960	82,356	13,490	1,971,152	930,133	554,347	375,786
1969	51,909	10,148	2,771,886	979,000	730,978	244,732
1978	43,075	9,461	6,314,665	1,861,266	1,325,936	535,330
1987	37,743	8,416	8,263,226	2,441,860	1,740,508	701,352
1997	31,757	7,254	9,102,191	2,834,512	1,834,095	1,000,417
2007	36,352	7,147	16,322,411	4,418,634	2,856,706	1,561,927

Empire State
Council of Agricultural Organizations

2009 MEMBERS

Dairy Farmers of America Northeast Council

Dairylea Cooperative, Inc.

Empire State Potato Growers, Inc.

Farm Credit Associations of New York

New York Apple Association, Inc.

New York Beef Producers Association

New York Corn Growers Association

New York Farm Bureau, Inc.

New York Poultry Association

New York Wine & Grape Foundation

New York State Agricultural Society

New York State Arborists - ISA Chapter, Inc.

New York State Flower Industries, Inc.

New York State Grange

New York State Horse Council

New York State Horticultural Society

New York State Maple Producers Association, Inc.

New York State Nursery and Landscape Association, Inc.

New York State Seed Association

New York State Turfgrass Association, Inc.

New York State Vegetable Growers Association, Inc.

New York State Veterinary Medical Society

Northeast Ag and Feed Alliance, Inc.

Northeast Cooperative Council

Northeast Dairy Foods Association, Inc.

Upstate Niagara Cooperative, Inc.

New York ● 1790

Canada

Lake Ontario

Lake Erie

CLINTON

VT

WASHINGTON

ALBANY

MONTGOMERY

COLUMBIA

MA

DUTCHESS

CT

ONTARIO

INDIAN LANDS

ULSTER

ORANGE

WESTCHESTER

SUFFOLK

QUEENS

KINGS

NEW YORK

RICHMOND

NJ

PA

Atlantic Ocean

CENSUS AVAILABILITY

Federal census extant for all counties.

BLACK = 1790 BOUNDARIES

100 MILES

0   25   50   75

25

1   2   3   4   5   6   7   8   9   10   11   12

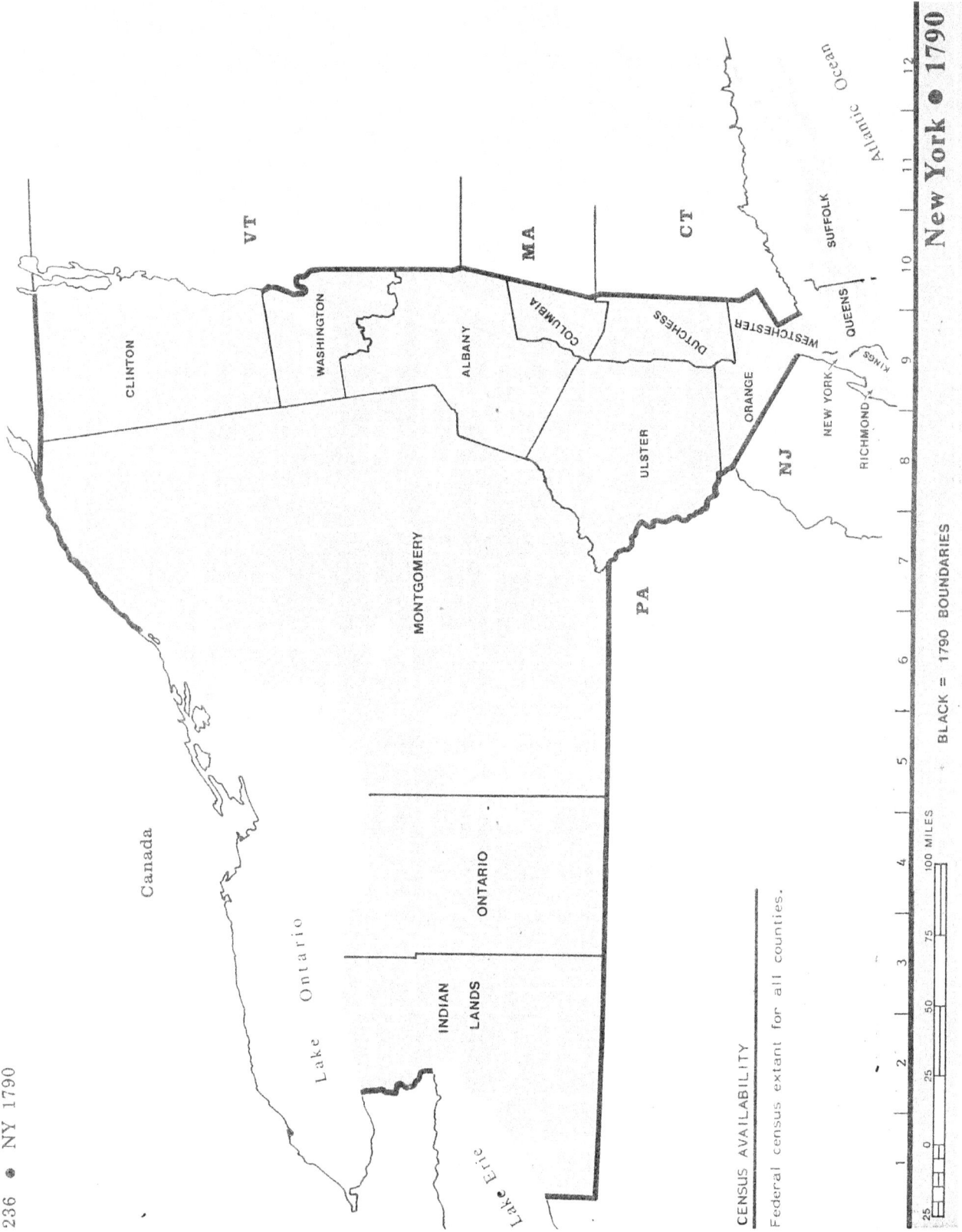

MAP GUIDE TO THE U.S. FEDERAL CENSUSES, 1790-1920 by William Thorndale and William Dollarhide. Copyright 1987, all rights reserved.

Reproduced with permission of authors.

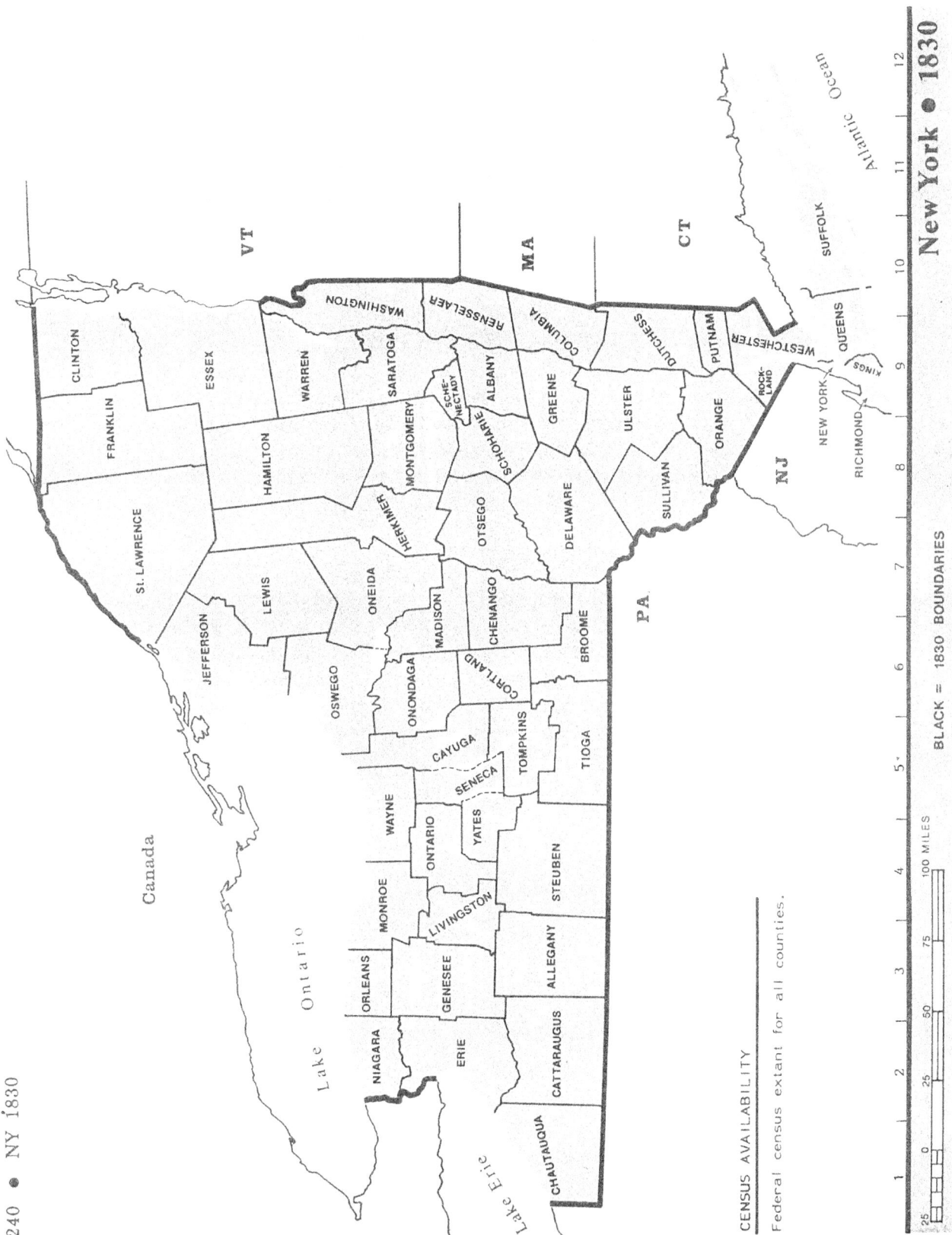

New York • 1830

CENSUS AVAILABILITY

Federal census extant for all counties.

BLACK = 1830 BOUNDARIES

100 MILES

MAP GUIDE TO THE U.S. FEDERAL CENSUSES, 1790-1920 by William Thorndale and William Dollarhide. Copyright 1987, all rights reserved.

Reproduced with permission of authors.

List of Counties in New York
FROM WIKIPEDIA, THE FREE ENCYCLOPEDIA

There are 62 counties in the State of New York. The first twelve counties in New York were created immediately after the British annexation of the Dutch colony of New Amsterdam, although two of these counties have since been abolished. The most recent county formation in New York was in 1912, when Bronx County was created from the portions of New York City that had been annexed from Westchester County. [1] New York's counties are named for a variety of Native American words, British provinces, cities, and royalty, early American statesmen and generals, and state politicians.[2]

Five of New York's counties are co-extensive with the five boroughs of New York City and do not have functioning county governments, except for a few borough officials. New York City is considered the county seat of these five counties: New York County (Manhattan), Kings County (Brooklyn), Bronx County (The Bronx), Richmond County (Staten Island), and Queens County (Queens). Because each borough has a separate main post office (and Queens has four), the county seats of the five boroughs are often stated in terms of those main post offices: New York (Manhattan), Brooklyn, Bronx, Staten Island, and Jamaica (Queens), NY. However, the communities served by those main post offices are all within the city limits of New York. In contrast to other counties of New York state, the powers of the five boroughs of New York City are very limited, and in nearly all respects subordinate to the city's.[3]

The FIPS county code is the five-digit Federal Information Processing Standard (FIPS) code which uniquely identifies counties and county equivalents in the United States. The three-digit number is unique to each individual county within a state, but to be unique within the entire United States, it must be prefixed by the state code. This means that, for example, while Albany County, New York is 001, Addison County, Vermont and Alachua County, Florida are also 001. To uniquely identify Albany County, New York, one must use the state code of 36 plus the county code of 001; therefore, the unique nationwide identifier for Albany County, New York is 36001. The links in the column FIPS County Code are to the Census Bureau Info page for that county.[4]

Alphabetical List of Counties

County	FIPS Code [4]	County Seat [5]	Created [5]	Formed from [1]	Named for [2]	Population [5]	Area [5]	Map
Albany	001	Albany	1683	One of 12 original counties created in the New York colony	King James II of England (1633–1701), who was Duke of York and Albany before ascending the throne of England, Duke of Albany being his Scottish title	294,565	533 sq mi (1,380 km²)	
Allegany	003	Belmont	1806	Genesee County	A variant spelling of the Allegheny River	49,927	1,034 sq mi (2,678 km²)	
Bronx	005	New York City (Coextensive with The Bronx)	1914[6]	New York County	Jonas Bronck (1600?–1643), an early settler of the Dutch colony of New Amsterdam	1,332,650	57.43 sq mi (149 km²)	
Broome	007	Binghamton	1806	Tioga County	John Broome (1738–1810), fourth Lieutenant Governor of New York	200,536	715 sq mi (1,852 km²)	

Four Hundred Years of Agricultural Change in the Empire State

Alphabetical List of Counties

County	FIPS Code [4]	County Seat [5]	Created [5]	Formed from [1]	Named for [2]	Population [5]	Area [5]	Map
Cattaraugus	009	Little Valley	1808	Genesee County	A Seneca word meaning "bad smelling banks," referring to the odor of natural gas which leaked from local rock formations	83,955	1,310 sq mi (3,393 km²)	
Cayuga	011	Auburn	1799	Onondaga County	The Cayuga tribe of Native Americans	81,963	864 sq mi (2,238 km²)	
Chautauqua	013	Mayville	1808	Genesee County	A Seneca word meaning "where the fish was taken out"	136,409	1,500 sq mi (3,885 km²)	
Chemung	015	Elmira	1836	Tioga County	A Lenape word meaning "big horn", which was the name of a local Native American village	91,070	410.81 sq mi (1,064 km²)	
Chenango	017	Norwich	1798	Tioga County and Herkimer County	An Onondaga word meaning "large bull-thistle"	51,401	898.85 sq mi (2,328 km²)	

Alphabetical List of Counties

County	FIPS Code [4]	County Seat [5]	Created [5]	Formed from [1]	Named for [2]	Population [5]	Area [5]	Map
Clinton	019	Plattsburgh	1788	Washington County	George Clinton (1739–1812), fourth Vice President of the United States and first and third Governor of New York	79,894	1,118 sq mi (2,896 km^2)	
Columbia	021	Hudson	1786	Albany County	Christopher Columbus (1451–1506), the European explorer	63,094	648 sq mi (1,678 km^2)	
Cortland	023	Cortland	1808	Onondaga County	Pierre Van Cortlandt (1721–1814), first Lieutenant Governor of New York	48,599	502 sq mi (1,300 km^2)	
Delaware	025	Delhi	1797	Otsego County and Ulster County	Thomas West, 3rd Baron De La Warr (1577–1618), an early colonial leader in Virginia	48,055	1,468 sq mi (3,802 km^2)	
Dutchess	027	Poughkeepsie	1683	One of 12 original counties created in the New York colony	Lady Anne Hyde (1637–1671), Duchess of York and wife of King James II of England	295,146	825 sq mi (2,137 km^2)	

Alphabetical List of Counties

County	FIPS Code [4]	County Seat [5]	Created [5]	Formed from [1]	Named for [2]	Population [5]	Area [5]	Map
Erie	029	Buffalo	1821	Niagara County	The Erie tribe of Native Americans	950,265	1,227 sq mi (3,178 km²)	
Essex	031	Elizabethtown	1799	Clinton County	The county of Essex in England	38,851	1,916 sq mi (4,962 km²)	
Franklin	033	Malone	1808	Clinton County	Benjamin Franklin (1706–1790), the early American printer, scientist, and statesman	51,134	1,697 sq mi (4,395 km²)	
Fulton	035	Johnstown	1838	Montgomery County	Robert Fulton (1765–1815), inventor of the steamship	55,073	533 sq mi (1,380 km²)	
Genesee	037	Batavia	1802	Ontario County	A Seneca phrase meaning "good valley"	60,370	495 sq mi (1,282 km²)	

Alphabetical List of Counties

County	FIPS Code [4]	County Seat [5]	Created [5]	Formed from [1]	Named for [2]	Population [5]	Area [5]	Map
Greene	039	Catskill	1800	Albany County and Ulster County	Nathanael Greene (1742–1786), the American Revolutionary War general	48,195	658 sq mi (1,704 km^2)	
Hamilton	041	Lake Pleasant	1816	Montgomery County	Alexander Hamilton (1755–1804), the early American political theorist and first United States Secretary of the Treasury	5,379	1,808 sq mi (4,683 km^2)	
Herkimer	043	Herkimer	1791	Montgomery County	Nicholas Herkimer (1728–1777), the American Revolutionary War general	64,427	1,458 sq mi (3,776 km^2)	
Jefferson	045	Watertown	1805	Oneida County	Thomas Jefferson (1743–1826), the early American statesman, author of the Declaration of Independence, and third President of the United States	111,738	1,857 sq mi (4,810 km^2)	
Kings	047	New York City (Coextensive with Brooklyn)	1683	One of 12 original counties created in the New York colony	King Charles II of England (1630–1685)	2,465,326	96.9 sq mi (251 km^2)	

Alphabetical List of Counties

County	FIPS Code [4]	County Seat [5]	Created [5]	Formed from [1]	Named for [2]	Population [5]	Area [5]	Map
Lewis	049	Lowville	1805	Oneida County	Morgan Lewis (1754–1844), the fourth Governor of New York	26,944	1,290 sq mi (3,341 km²)	
Livingston	051	Geneseo	1821	Genesee County and Ontario County	Robert Livingston (1746–1813), the early American statesman and New York delegate to the Continental Congress	64,328	640 sq mi (1,658 km²)	
Madison	053	Wampsville	1806	Chenango County	James Madison (1751–1836), the early American statesman, principal author of the Constitution of the United States, and fourth President of the United States	69,441	662 sq mi (1,715 km²)	
Monroe	055	Rochester	1821	Genesee County and Ontario County	James Monroe (1758–1831), the early American statesman and fifth President of the United States	735,343	1,366 sq mi (3,538 km²)	
Montgomery	057	Fonda	1772	Albany County	Originally Tryon County after colonial governor William Tryon (1729–1788), renamed after the American Revolutionary War general Richard Montgomery (1738–1775) in 1784	49,708	410 sq mi (1,062 km²)	

Alphabetical List of Counties

County	FIPS Code [4]	County Seat [5]	Created [5]	Formed from [1]	Named for [2]	Population [5]	Area [5]	Map
Nassau	059	Mineola	1899	Queens County	William of Orange-Nassau (1650–1702), who would become King William III of England	1,334,544	453 sq mi (1,173 km^2)	
New York	061	New York City (Coextensive with Manhattan)	1683	One of 12 original counties created in the New York colony	King James II of England (1633–1701), who was Duke of York and Albany before he ascended the throne of England, Duke of York being his English title	1,537,195	33.77 sq mi (87 km^2)	
Niagara	063	Lockport	1808	Genesee County	An Iroquoian word perhaps meaning "a neck" between two bodies of water, "thunder of waters," or "bisected bottom land"	219,846	1,140 sq mi (2,953 km^2)	
Oneida	065	Utica	1798	Herkimer County	The Oneida tribe of Native Americans	235,469	1,213 sq mi (3,142 km^2)	
Onondaga	067	Syracuse	1792	Herkimer County	The Onondaga tribe of Native Americans	458,336	806 sq mi (2,088 km^2)	

Alphabetical List of Counties

County	FIPS Code [4]	County Seat [5]	Created [5]	Formed from [1]	Named for [2]	Population [5]	Area [5]	Map
Ontario	069	Canandaigua	1789	Montgomery County	An Iroquoian word meaning "beautiful lake"	100,224	662 sq mi (1,715 km²)	
Orange	071	Goshen	1683	One of 12 original counties created in the New York colony	William of Orange-Nassau (1650–1702), who would become King William III of England	341,367	839 sq mi (2,173 km²)	
Orleans	073	Albion	1824	Genesee County	The French Royal House of Orleans	44,171	817 sq mi (2,116 km²)	
Oswego	075	Oswego	1816	Oneida County and Onondaga County	The Oswego River, from an Iroquoian word meaning "the outpouring", referring to the mouth of the river	122,377	1,312 sq mi (3,398 km²)	
Otsego	077	Cooperstown	1791	Montgomery County	A Native American word meaning "place of the rock"	61,676	1,003 sq mi (2,598 km²)	

Alphabetical List of Counties

County	FIPS Code [4]	County Seat [5]	Created [5]	Formed from [1]	Named for [2]	Population [5]	Area [5]	Map
Putnam	079	Carmel	1812	Dutchess County	Israel Putnam (1718–1790), an American Revolutionary War general	95,745	246 sq mi (637 km^2)	
Queens	081	New York City (Coextensive with Queens)	1683	One of 12 original counties created in the New York colony	Catherine of Braganza (1638–1705), Queen of England and wife of King Charles II of England	2,229,379	178.28 sq mi (462 km^2)	
Rensselaer	083	Troy	1791	Albany County	Kiliaen van Rensselaer (1585–1643), the early landholder in the Dutch New Amsterdam colony	152,538	665 sq mi (1,722 km^2)	
Richmond	085	New York City (Coextensive with Staten Island)	1683	One of 12 original counties created in the New York colony	Charles Lennox, 1st Duke of Richmond (1672–1723), the illegitimate son of King Charles II of England	443,728	102.5 sq mi (265 km^2)	
Rockland	087	New City	1798	Orange County	Early settlers' description of terrain as "rocky land"	286,753	199 sq mi (515 km^2)	

Alphabetical List of Counties

County	FIPS Code [4]	County Seat [5]	Created [5]	Formed from [1]	Named for [2]	Population [5]	Area [5]	Map
St. Lawrence	089	Canton	1802	Clinton County, Herkimer County, and Montgomery County	The St Lawrence River, which forms the northern border of the county and New York State	111,931	2,821 sq mi (7,306 km²)	
Saratoga	091	Ballston Spa	1791	Albany County	A corruption of a Native American word meaning "the hill beside the river"	200,635	844 sq mi (2,186 km²)	
Schenectady	093	Schenectady	1809	Albany County	A Mohawk word meaning "on the other side of the pine lands"	146,555	210 sq mi (544 km²)	
Schoharie	095	Schoharie	1795	Albany County and Otsego County	A Mohawk word meaning "floating driftwood"	31,582	626 sq mi (1,621 km²)	
Schuyler	097	Watkins Glen	1854	Chemung County, Steuben County, and Tompkins County	Philip Schuyler (1733–1804), the American Revolutionary War general and Senator from New York	19,224	342 sq mi (886 km²)	

Alphabetical List of Counties

County	FIPS Code [4]	County Seat [5]	Created [5]	Formed from [1]	Named for [2]	Population [5]	Area [5]	Map
Seneca	099	Ovid and Waterloo	1804	Cayuga County	The Seneca tribe of Native Americans	33,342	325 sq mi (842 km^2)	
Steuben	101	Bath	1796	Ontario County	Friedrich Wilhelm von Steuben (1730–1794), the Prussian general who assisted the Continental Army during the American Revolutionary War	98,726	1,404 sq mi (3,636 km^2)	
Suffolk	103	Riverhead	1683	One of 12 original counties created in the New York colony	The county of Suffolk in England	1,419,369	2,373 sq mi (6,146 km^2)	
Sullivan	105	Monticello	1809	Ulster County	John Sullivan (1740–1795), an American Revolutionary War general	73,966	997 sq mi (2,582 km^2)	
Tioga	107	Owego	1791	Montgomery County	A Native American word meaning "at the forks," describing a meeting place	51,784	523 sq mi (1,355 km^2)	

Alphabetical List of Counties

County	FIPS Code [4]	County Seat [5]	Created [5]	Formed from [1]	Named for [2]	Population [5]	Area [5]	Map
Tompkins	109	Ithaca	1817	Cayuga County and Seneca County	Daniel D. Tompkins (1774–1825), the 6th Vice President of the United States	96,501	476 sq mi (1,233 km²)	
Ulster	111	Kingston	1683	One of 12 original counties created in the New York colony	The Irish province of Ulster, then an earldom of the Duke of York, later King James II of England	177,749	1,161 sq mi (3,007 km²)	
Warren	113	Queensbury	1813	Washington County	Joseph Warren (1741–1775), the early American patriot and American Revolutionary War general	63,303	870 sq mi (2,253 km²)	
Washington	115	Fort Edward	1772	Albany County	Originally Charlotte County, renamed in 1784 after George Washington (1732–1799), the American Revolutionary War general and first President of the United States	61,042	846 sq mi (2,191 km²)	
Wayne	117	Lyons	1823	Ontario County and Seneca County	General Anthony Wayne (1745–1796), the American Revolutionary War general	93,765	1,384 sq mi (3,585 km²)	

Alphabetical List of Counties

County	FIPS Code [4]	County Seat [5]	Created [5]	Formed from [1]	Named for [2]	Population [5]	Area [5]	Map
Westchester	119	White Plains	1683	One of 12 original counties created in the New York colony	The city of Chester in England	923,459	500 sq mi (1,295 km²)	
Wyoming	121	Warsaw	1841	Genesee County	A modification of a word from the language of the Lenape tribe of Native Americans meaning "broad bottom lands"	43,424	596 sq mi (1,544 km²)	
Yates	123	Penn Yan	1823	Ontario County and Steuben County	Joseph C. Yates (1768–1837), eighth Governor of New York	24,621	376 sq mi (974 km²)	

REFERENCES

1. a b c d e "New York Formation Maps". *Genealog, Inc.* Retrieved on 2008-01-20.
2. a b Beatty, Michael (2001). *County Name Origins of the United States.* McFarland Press. ISBN 0786410256.
3. Benjamin, Gerald, Richard P. Nathan (1990). *Regionalism and Realism: A Study of Government in the New York Metropolitan Area.* Brookings Institute. pp. 59.
4. a b "EPA County FIPS Code Listing". *US Environmental Protection Agency.* Retrieved on 2007-07-24.
5. a b c d "NACo - Find a county". *National Association of Counties.* Retrieved on 2008-01-20.
6. Legislation splitting off Bronx from Westchester was enacted in 1912 with an effective date of January 1, 1914. See McCarthy, Thomas C.. "A 5-Borough Centennial Preface for the Katharine Bement Davis Mini-History". *New York City Department of Corrections.* Retrieved on 2008-01-25.
7. Lynch, Mike (2007-10-30). "North Elba Supervisor Candidate Debate", *Plattsburgh Press Republican.* Retrieved on 20 January 2008.
8. Healy, Patrick (2004-02-11). "Growth Pains and Clout Heading East in Suffolk", *The New York Times.* Retrieved on 20 January 2008.

Bibliography

A History of Agriculture in the State of New York, Ulysses Prentiss Hedrick

The Sullivan-Clinton Campaign in 1779, State University of the State of New York 1929

The Old New York Frontier, Francis Whiting Halsey

American Commonwealths, Ellis H. Roberts

Colonial New York, Michael Kammen

Corn in the Culture of the Mohawk Iroquois, Daniel K. Onion

The Science Behind the Three Sisters Mound System, Jane Mt Pleasant

Historical Sketch of American Agriculture in Bailey's Cyclopedia, T. N. Carver

Historical Sketch of American Agriculture in Bailey's Cyclopedia, G. K. Holmes

From the New Netherland Institute, Albany, Newsday. com

The American Revolution in New York, New York Division of Archives & History

The Golden Age of Homespun, Jared van Wagenen, Jr.

The Return of Agricultural Lands to Forest, Bernard F. Stanton & Nelson L. Bills

Old Mohawk Turnpike Book, Charles B. Knox Gelatin Company

Learning from the 19th Century; Private Toll Roads, Daniel B. Klein & Gordon J. Fielding

The Cato Journal, Fall 2002, Clifford F. Thies

History of American Agriculture, Farm Machinery and Technology, USDA Publication

A History of the Canning and Freezing Industry in New York State, The Associated New York State Food Processors

New York's Sugar Beet Fiasco, Gould P. Colman & Jerry D. Stockdale

A Guide to the Wilderness, William Cooper

The Development of Central and Western New York, Clayton Mau

From the Laws and Ordinances of New Netherland, Edmund Bailey O'Callaghan

History of the New York State Grange, L. Ray Alexander

The History of the New York State Grange 1934-1960, Elizabeth L. Arthur

History of New York Farm Bureau, New York Farm Bureau, Inc.

Education and Agriculture, Gould P. Colman

Dairylea 100 Years of Service, Dairylea

A Century's Progress in Agricultural Education, Cornelius A. Betten

The Contract Colleges of Cornell University, Malcolm Carron

The People's Colleges, Ruby Green Smith

New York State Census, 1825, 1835, 1845, 1855, 1865, 1875

United States Agricultural Census, 1850-2009

United States Census, 1790-1840

New York Agricultural Statistics, New York State Department of Ag & Mkts

Index

M

Madison 50, 72, 115, 186, 187, 188, 189, 190, 195, 196, 198, 199, 212
Manhattan 5, 9, 206, 213
A.R. Mann 160
Maple Lawn Dairy 85
maple sugar 4, 12, 25, 113, 114, 121
maple syrup 4, 113, 114, 121, 122, 173
Massachusetts 7, 14, 23, 105, 130, 191, 192, 193, 194, 197
Cyrus McCormick 47
mechanization 25, 43, 50, 52, 54, 80
Merino sheep 23
Merrell and Soule 93
Metropolitan Cooperative Milk Producers Bargaining Agency 151
Mike Weaver Drain Tile Museum 125
Military Tract 14
milk 11, 23, 41, 52, 55, 58, 59, 68, 75, 76, 77, 78, 79, 80, 81, 82, 83, 84, 85, 93, 110, 120, 121, 122, 137, 147, 148, 149, 150, 151, 152, 153, 154, 160, 168, 173, 175, 176, 199
Ebenezer Miller 7
mink 86
Mohawk 1, 2, 3, 8, 9, 16, 21, 27, 29, 30, 33, 38, 39, 44, 50, 120, 139, 141, 216, 221
Mohawk and Hudson 16, 30, 38, 39, 141
Mohawk and Hudson Turnpike 30
Mohawk Turnpike 29, 30, 50, 221
Mohawk Valley 1, 8, 9, 21, 30, 44
Monroe 70, 107, 110, 113, 136, 164, 189, 190, 195, 196, 198, 199, 212
Montauk 7
Montezuma 30, 116
Montgomery 75, 130, 186, 187, 188, 189, 190, 195, 196, 198, 199, 210, 211, 212, 214, 216, 217
Robert Morrison 14

N

Nassau 27, 62, 110, 189, 190, 196, 198, 199, 213, 214
National Grange of the Patrons of Husbandry 135

National Grape Cooperative Association, Inc. 107
Native American iii, 1, 2, 3, 12, 13, 21, 206, 208, 214, 216, 217
natural v, 16, 29, 58, 60, 83, 84, 127, 128, 141, 155, 163, 168, 170, 174, 175, 208
Natural Resources Conservation Service 169, 170
Netherlands 5, 6
New England 5, 10, 13, 16, 36, 38, 41, 70, 139, 141, 159
New Netherlands 5, 6
New York and Albany Post Road 29
New York Canned Goods Packers' Association 95
New York Central railroad 39
New York City 5, 7, 8, 12, 13, 16, 18, 19, 27, 29, 30, 33, 34, 36, 39, 40, 41, 60, 66, 93, 132, 141, 143, 145, 146, 155, 169, 206, 207, 211, 213, 215, 219
New York Farm Bureau 137, 138, 203, 221
New York Farm Viability Institute 140
New York Market Order 151
New York State Agricultural Experiment Station 157
New York State Agricultural Society vi, 54, 70, 125, 129, 130, 131, 132, 133, 134, 156, 159, 168, 203
New York State Apple Growers Association 139
New York State Artificial Breeders Cooperative 83
New York State Fair 131, 132, 134, 168
New York Wine & Grape Foundation 140, 203
Niagara 21, 101, 107, 144, 152, 153, 186, 187, 188, 189, 190, 195, 196, 198, 199, 203, 210, 213
Northeast Dairy Herd Improvement Association 84, 153
Northern Traveler 36
nursery 15, 101, 103, 110, 132, 192, 203
nuts 4, 101, 109
NYS Horticultural Society 135, 139
NYS Vegetable Growers Association 135, 140

O

O-AT-KA 152
oats 23, 59, 69, 70, 193
Dan O'Hara 133
Ohio 14, 30, 32, 38, 40, 76, 105, 137, 191, 192, 193, 194, 197
Ohio Trail 30
Oneida 2, 33, 36, 75, 78, 79, 115, 122, 124, 130, 186, 187, 188, 189, 190, 195, 196, 198, 199, 211, 212, 213, 214
onions 91, 123
Onondaga iv, 2, 14, 17, 22, 34, 49, 61, 62, 66, 67, 68, 70, 72, 84, 88, 89, 90, 94, 97, 108, 114, 115, 117, 121, 124, 125, 132, 141, 147, 148, 151, 152, 160, 161, 186, 187, 188, 189, 190, 195, 196, 198, 199, 208, 209, 213, 214
Onondaga Milk Association 147
Ontario 13, 21, 26, 32, 33, 76, 101, 107, 142, 186, 187, 188, 189, 190, 195, 196, 198, 199, 210, 212, 214, 217, 218, 219
Orange 5, 6, 8, 76, 78, 79, 91, 109, 110, 126, 149, 186, 187, 188, 189, 190, 195, 196, 198, 199, 213, 214, 215
organic 3, 56, 58, 60, 68, 73, 84, 89, 90, 92, 106, 108, 152, 161, 163, 164, 168, 173, 174, 175
Organic Valley Cooperative 152
Orleans 36, 91, 101, 105, 126, 189, 190, 195, 196, 198, 199, 214
D.M. Osborne Company 47
Oswego 2, 14, 21, 27, 32, 36, 38, 101, 109, 189, 190, 195, 196, 198, 199, 214
Otsego 23, 26, 75, 115, 130, 177, 181, 186, 187, 188, 189, 190, 195, 196, 198, 199, 209, 214, 216
Oyster Bay Historical Society 8

P

Palatine Bridge 8
Palatines 8, 9
pasture v, 7, 20, 59, 65, 66, 68, 76, 88, 89, 111, 117, 124, 126, 176
Patent Office 44
patents 8, 23, 44
patroon 6
peaches 107, 108

THE ASTOUNDING UFO SECRETS
Of James W. Moseley

A SPECIAL TRIBUTE TO THE EDITOR OF SAUCER SMEAR AND THE COURT JESTER OF UFOLOGY

Includes The Full Text Of UFO Crash Secrets At Wright Patterson Air Force Base

Edited By Timothy Green Beckley

GLOBAL COMMUNICATIONS
EX LIBRIS

POST OFFICE BOX 753
NEW BRUNSWICK, NJ 08903